A Bayesian Introduction to Fish Population Analysis

Fisheries science is an applied field. A biologist working on a depleted population might need to forecast the population trajectory or examine which factors are limiting recovery. For an exploited population, a biologist might investigate whether the fishery or the stock would be improved through harvest regulation, such as a quota or size limit. Addressing those sorts of questions generally involves estimating parameters such as population size or survival rate. An essential step is characterizing the degree of uncertainty in results, which can be substantial for fisheries models. Bayesian methods and current software provide a flexible and powerful framework for addressing these challenges. Models developed using Bayesian software can be tailored to the unique aspects of each field study.

A Bayesian Introduction to Fish Population Analysis is aimed at advanced undergraduate and graduate students as well as working professionals interested in a hands-on introduction to Bayesian approaches for fitting fisheries models. Chapters address key aspects of population dynamics: abundance, mortality, growth, and recruitment. The book includes complete R code for simulating each study design and JAGS Bayesian code for model fitting; code files are also available online. No prior knowledge of R or JAGS is assumed, and new commands are introduced gradually through the sequence of examples. There is emphasis throughout the book on how to vary simulation settings to develop intuition about fisheries models (e.g., how many fish should be tagged in order to obtain usefully precise results). Additional topics include development of integrated population models, model checking, model selection, and uninformative and informative prior distributions.

Key features include:
- Early chapters focus on ecologically relevant probability distributions (e.g., binomial, Poisson, normal, lognormal) as a foundation for the applied fisheries chapters.
- Subsequent chapters demonstrate Bayesian approaches for estimating abundance, mortality, growth, and recruitment.
- Full open-source code is provided for simulation and plotting using R and Bayesian model fitting using JAGS.
- Simulation is used to gain a deeper understanding of analytical methods and for planning field studies.

Joseph E. Hightower is a professor emeritus in the Department of Applied Ecology at NC State University. His research interests focus on fish population dynamics, especially field and analytical methods for estimating population parameters. His primary teaching role was a graduate course in quantitative fisheries management, including Bayesian methods that serve as the foundation for this book.

Chapman & Hall/CRC
Applied Environmental Series

Series Editors

Douglas Nychka, *Colorado School of Mines*
Alexandra Schmidt, *Universidade Federal do Rio de Janero*
Richard L. Smith, *University of North Carolina*
Lance A. Waller, *Emory University*

Recently Published Titles

Sampling Techniques for Forest Inventories
Daniel Mandallaz

Statistics for Environmental Science and Management
Second Edition
Bryan F.J. Manly

Statistical Geoinformatics for Human Environment Interface
Wayne L. Myers, Ganapati P. Patil

Introduction to Hierarchical Bayesian Modeling for Ecological Data
Eric Parent, Etienne Rivot

Handbook of Spatial Point-Pattern Analysis in Ecology
Thorsten Wiegand, Kirk A. Moloney

Introduction to Ecological Sampling
Bryan F.J. Manly, Jorge A. Navarro Alberto

Future Sustainable Ecosystems: Complexity, Risk, and Uncertainty
Nathaniel K Newlands

Environmental and Ecological Statistics with R
Second Edition
Song S. Qian

Statistical Methods for Field and Laboratory Studies in Behavioral Ecology
Scott Pardo, Michael Pardo

Biometry for Forestry and Environmental Data with Examples in R
Lauri Mehtätalo, Juha Lappi

Bayesian Applications in Environmental and Ecological Studies with R and Stan
Song S. Qian, Mark R. DuFour, Ibrahim Alameddine

Spatio-Temporal Models for Ecologists
James Thorson and Kasper Kristensen

Spatial Linear Models for Environmental Data
Dale L. Zimmerman and Jay Ver Hoef

A Bayesian Introduction to Fish Population Analysis
Joseph E. Hightower

For more information about this series, please visit: https://www.crcpress.com/Chapman--HallCRC-Applied-Environmental-Statistics/book-series/ CRCAPPENVSTA

A Bayesian Introduction to Fish Population Analysis

Joseph E. Hightower

CRC Press
Taylor & Francis Group
Boca Raton London New York

CRC Press is an imprint of the
Taylor & Francis Group, an **informa** business

A CHAPMAN & HALL BOOK

First edition published 2026
by CRC Press
2385 NW Executive Center Drive, Suite 320, Boca Raton FL 33431

and by CRC Press
4 Park Square, Milton Park, Abingdon, Oxon, OX14 4RN

CRC Press is an imprint of Taylor & Francis Group, LLC

© 2026 Joseph E. Hightower

Library of Congress CataloginginPublication Data
Names: Hightower, Joseph E., author.
Title: A Bayesian introduction to fish population analysis / Joseph E
Hightower.
Description: First edition. | Boca Raton, FL : Taylor and Francis, 2026. |
Includes bibliographical references and index. |
Summary: "Fisheries science is an applied field. A biologist working on a depleted
population might need to forecast the population trajectory or examine
which factors are limiting recovery. For an exploited population, a
biologist might investigate whether the fishery or the stock would b
improved through harvest regulation, such as a quota or size limit.
Addressing those sorts of questions generally involves estimating
parameters such as population size or survival rate. An essential step
is characterizing the degree of uncertainty in results, which can be
substantial for fisheries models. Bayesian methods and current software
provide a flexible and powerful framework for addressing these
challenges. Models developed using Bayesian software can be tailored to
the unique aspects of each field study. A Bayesian Introduction to Fish
Population Analysis is aimed at advanced undergraduate and graduate
students as well as working professionals interested in a hands-on
introduction to Bayesian approaches for fitting fisheries models.
Chapters address key aspects of population dynamics: abundance,
mortality, growth, and recruitment. The book includes complete R code
for simulating each study design and JAGS Bayesian code for model
fitting; code files are also available online. No prior knowledge of R
or JAGS is assumed and new commands are introduced gradually through the
sequence of examples. There is emphasis throughout the book on how to
vary simulation settings to develop intuition about fisheries models
(e.g., how many fish should be tagged in order to obtain usefully
precise results). Additional topics include development of integrated
population models, model checking, model selection, and uninformative
and informative prior distributions"-- Provided by publisher.
Identifiers: LCCN 2025005405 (print) | LCCN 2025005406 (ebook) | ISBN
9781032833552 (hardback) | ISBN 9781032840901 (paperback) | ISBN
9781003511151 (ebook)
Subjects: LCSH: Fish stock assessment--Mathematical models. | Fishery
sciences--Mathematical models. | Fishery management. | Bayesian
statistical decision theory.

ISBN: 978-1-032-83355-2 (hbk)
ISBN: 978-1-032-84090-1 (pbk)
ISBN: 978-1-003-51115-1 (ebk)

DOI: 10.1201/9781003511151

Typeset in Latin Modern font
by KnowledgeWorks Global Ltd.

To my former graduate students, whose field studies led to rewarding collaborations on Bayesian fisheries models.

Contents

Preface

This book is based in large part on material I developed while teaching (1991-2014) at NC State University. My hope is that the book will be a bridge between traditional fisheries analytical methods and Bayesian approaches that offer many advantages in ecological modeling. Many other texts address fish population dynamics and quantitative management (e.g., Hilborn and Walters, 1992; Quinn and Deriso, 1999); here the focus is strictly on Bayesian approaches for fitting fisheries models. The book might be useful as an upper-level undergraduate or early graduate text, or for a working fisheries biologist interested in a hands-on introduction to Bayesian methods.

The general approach for this book follows that used by Kéry (2010) and Kéry and Schaub (2012), in that sample data for each Bayesian analysis are produced by simulation. Analyzing simulated data is helpful for learning because the true parameter values are known. Simulating a field study also provides the flexibility to vary study parameters such as the number of samples or sampling variation. Running the simulation and analysis code multiple times under different settings will aid in understanding the methods and their limitations. This brings up a general point about this book: reading about an analysis is not enough – it is essential to run the code in order to understand the methods. Exercises included at the end of each chapter should also aid in understanding the strengths and weaknesses of each analysis; solutions are available as an online file[1]. The book includes an appendix with empirical examples from a subset of analyses to illustrate the changes in code required.

Software information and conventions

This book has been prepared using the **knitr** (Xie, 2025b) and **bookdown** (Xie, 2025a) packages. Package names (e.g., **bookdown**) will be indicated by bold text; typewriter font will indicate inline code (e.g., `x^y`) or a function name (followed by parentheses, e.g., `mean()`). My R session information is shown below:

---

[1]https://github.com/jhncsu/BayesfishFiles/blob/main/Exercises.pdf

```
xfun::session_info()
```

```
## R version 4.4.2 (2024-10-31 ucrt)
## Platform: x86_64-w64-mingw32/x64
## Running under: Windows 11 x64 (build 26100)
##
## Locale:
##   LC_COLLATE=English_United States.utf8
##   LC_CTYPE=English_United States.utf8
##   LC_MONETARY=English_United States.utf8
##   LC_NUMERIC=C
##   LC_TIME=English_United States.utf8
##
## Package version:
##   base64enc_0.1.3    bookdown_0.42
##   bslib_0.9.0        cachem_1.1.0
##   cli_3.6.4          compiler_4.4.2
##   digest_0.6.37      evaluate_1.0.3
##   fastmap_1.2.0      fontawesome_0.5.3
##   fs_1.6.5           glue_1.8.0
##   graphics_4.4.2     grDevices_4.4.2
##   highr_0.11         htmltools_0.5.8.1
##   jquerylib_0.1.4    jsonlite_2.0.0
##   knitr_1.50         lifecycle_1.0.4
##   memoise_2.0.1      methods_4.4.2
##   mime_0.13          R6_2.6.1
##   rappdirs_0.3.3     rlang_1.1.5
##   rmarkdown_2.29     sass_0.4.9
##   stats_4.4.2        tinytex_0.56
##   tools_4.4.2        utils_4.4.2
##   xfun_0.52          yaml_2.3.10
```

Acknowledgments

My first exposure to Bayesian statistical methods was via Matthew Krachey, who took the time as a PhD student to teach a few informal sessions on WinBUGS. I soon came to agree with Kéry (2010) that the BUGS language "frees the modeler in you" and would be a valuable tool for research and especially teaching. I incorporated the BUGS language into my graduate fisheries course and soon into research population models. Other colleagues who provided valuable assistance on statistical methods include Beth Gardner, Ken Pollock, and Brian Reich. I also benefited from some of the many excellent reference books on Bayesian methods for ecology (e.g., McCarthy, 2007; Kéry, 2010; Kéry and Schaub, 2012).

Julie Harris, Jacob Krause, Paul Rudershausen and several anonymous reviewers made helpful comments on previous drafts. I have attempted to address those suggestions; any remaining errors in the book are mine.

About the Author

Joseph E. Hightower is Professor Emeritus in the Department of Applied Ecology at NC State University in Raleigh, North Carolina. Prior to retiring, his research focused on fish population dynamics and methods of estimating population parameters. Field studies primarily addressed the ecology of anadromous fishes including striped bass, American shad and sturgeon species.

1

Introduction

Survey a group of fishery biologists and many would say that they entered the field because of a love of the outdoors. They might also admit that, despite a preference for being outside, much of their work is done at a computer. There are mundane tasks like email correspondence, but of greater importance (and relevance for this book) are the steps in getting the information needed for effective management. Those steps include planning and conducting field studies, carrying out correct analyses, and presenting results in reports and presentations. A strong statistical foundation is essential for accomplishing these tasks. Otherwise, management ends up being a trial-and-error process that is inefficient and often ineffective. The purpose of this book is to provide an introduction to these quantitative methods.

The models that we will investigate have a statistical foundation, but our focus will be on parameter estimation rather than hypothesis testing. This is the opposite of most university statistics courses, but it reflects the reality of fishery management. For example, we might conduct a tagging study in order to estimate the exploitation rate, or the fraction of the legal-sized fish that are harvested in a year. We are not interested in a trivial hypothesis test ("Is the exploitation rate 0?") but just in getting a reliable estimate. If a substantial fraction of the population is being harvested, then the fishery manager might consider harvest restrictions such as a closed season, bag limit, or size limit. The tagging study could be done using red tags that have a reward of $100 and yellow tags that have a reward of $5. Anglers typically return high-reward tags at a higher rate than low-reward tags, so one part of the planning process would be to decide how many tags of each type to use. Once the study is complete, the model used for planning can often be modified for analysis of the tag-return data. One step in the process would be to determine whether the model needs separate parameters for the return rates of high- and low-reward tags. That could be viewed as a hypothesis-testing situation, but can also be thought of as model selection (simpler model with one rate versus a more complex model with two). Ultimately, we simply want to produce the best possible information about the effect of fishing on the population. This would include not only a point estimate but also measures of uncertainty that aid in decision-making.

Examples in the chapters that follow use Bayesian statistical methods to illustrate the steps in planning and analysis of field studies. Fisheries analyses (and university statistics courses) have traditionally been based on so-called frequentist statistical methods, that are based on the expected frequency of observing the data in hand, if the study were repeated many times (McCarthy, 2007). Until recently, an important advantage of frequentist methods was that parameter estimation, hypothesis testing, etc. could be implemented with commonly available computing resources (Dorazio, 2016). In contrast, Bayesian methods of statistical inference were not practical until fairly recently, when generic algorithms known as Markov Chain Monte Carlo (Section 4.3) became available (Dorazio, 2016). Another important factor in the increased adoption of Bayesian methods is the availability of public-domain software to fit the models (Kéry and Schaub, 2012). The earliest version was known as BUGS, for Bayesian inference Using Gibbs Sampling (Lunn et al., 2009), followed by a Windows version WinBUGS (Lunn et al., 2000), then an open version OpenBUGS (Lunn et al., 2009), as well as JAGS (Plummer, 2003) which was developed independently but uses essentially identical code. The BUGS language is a great tool both for learning and for getting work done. Rather than trying to force the data into a rigid black box for analysis, study-specific code can be produced that is readable and biologically meaningful. Another advantage of BUGS software is that it forces you to think about the biological and field-sampling processes that generated your data. I use JAGS in this book, but the code should work without modification in other versions of the BUGS family, and it could be converted to run under other Bayesian software such as NIMBLE (de Valpine et al., 2017) or Stan (Gelman et al., 2015). The JAGS code used in this book will be run from R (R Core Team, 2024), a general software environment that is briefly described in Chapter 2.

2

A Brief Introduction to R and RStudio

R (R Core Team, 2024) is a free software environment for data manipulation, statistical analyses, and graphics. It is easily extended through add-on software packages including some for commonly used fisheries models. There are a number of software interfaces for working in R and you should use the interface with which you are most comfortable. For this book, I use the free software RStudio[1], which can be run in the Windows, Macintosh, or Unix environment. There are many online resources for installing and learning R and RStudio, and even a recent book for fisheries analysis using R (Ogle, 2016). I assume in this book that R and RStudio have been installed. This chapter introduces a few basic R commands, but I primarily illustrate the use of R through code examples in the remaining chapters. I also describe some general methods for getting help and learning more about R.

As a first example using R, let's estimate population size through a two-sample mark-recapture study. Consider a pond with N=100 fish. If you take a first sample (n_1=50) and mark (tag) those fish, then half the fish in the pond are tagged. If you take a second sample (say n_2=30), how many tagged fish (m_2) would you expect in the second sample? That's right, 15! Of course, in practice, we don't know N, but we solve for it. We begin by setting the two proportions equal: $n_1/N = m_2/n_2$. Solving for N gives us our estimator: $\hat{N} = \frac{n_1*n_2}{m_2}$.

We can carry out the calculations using R code as follows. First, open a new RStudio window (menu commands File : New File : R Script). This Source window will contain your R script, or the sequence of R commands you intend to execute. Enter or copy and paste the following code into the Source window (from the online file[2] if using a print copy of the book):

```
setwd("Q:/My Drive/FishAnalysis") #Edit to show your working directory

# Two-sample population estimate
n1 <- 50 # Number caught and marked in first sample
n2 <- 30 # Number caught and examined for marks in second sample
```

---

[1]https://posit.co/products/open-source/rstudio/
[2]https://github.com/jhncsu/BayesfishFiles/blob/main/CodeForBook.txt

```
m2 <- 15 # Number of marked fish in second sample
N_hat <- n1*n2/m2 # Store estimate as variable N_hat
N_hat # Prints N_hat in Console window
```

Any text following a pound sign (#) is a comment, whether on a line by itself or following executable code. Comments do not affect the program's execution, but are extremely helpful in documenting your code. Taking the time to enter comments is always worthwhile, whether for your own information or when code is to be shared with others.

Next, position the cursor on the first line and execute one line at a time by clicking on the Run icon. (Note that if you hover the cursor over the Run icon without clicking, RStudio shows that Ctrl-Enter is a keyboard shortcut.) Running code one line at a time allows you to see the effect of each statement. The first line calls the setwd() function to establish the working directory for any files saved. Doing this routinely will ensure that files end up in the proper location. You can find out your working directory path by using an RStudio menu command (Session | Set Working Directory | Choose Directory). That allows you to navigate to the correct location, and the setwd() code will be printed in the Console window. You can then save that line of code for future use.

The next three executable lines assign (<-) values to variables, which have arbitrary names. Here, I have used conventional fisheries mark-recapture names for the sample sizes and number of recaptures, but more descriptive names could be used (e.g., Mark_Sample) for clarity. Each variable is a new object that will appear in the Environment window once that line of code has been run. The next line of code calculates the estimate, which is stored as N_hat. The final line prints the value of N_hat in the Console window. The population estimate (100) makes sense because you marked 50, and half the fish in the second sample were marked.

After you have run this code in RStudio, be sure to save the file. This is a good general practice – never create code from scratch when you can modify existing code! It cuts down on typing errors and allows you to build on existing code when doing new but similar things. The code could be from your prior work or from a journal article or book. The ability to share R and JAGS code is very helpful in learning and working efficiently. Saving the file to a working directory specified in the code helps in keeping together all files related to a specific project. Use a descriptive name for the code (R script) so that it can be readily located in the future.

Much of the work done in R is accomplished by using functions. Each function is a block of code that performs a specific task. Typing the name of an R function into the search box in the RStudio Help window is one quick way to get help (e.g., function defaults, arguments, examples). However, the R documentation is sometimes overly detailed and cryptic, so another good option is an online search (e.g., "R function max" or "how to load csv file R"). Another helpful alternative is the R Reference Card[3], which provides a brief summary of some of the more frequently used R commands. For example, the reference card lists a number of math functions such as log(x, base), which returns the log of x to the specified base. The statement log(x) returns the default natural-log of x. This is typical of R, in that functions have default arguments that need not be specified. (It can also be a bit dangerous if the default action is not what you intended!) Another important aspect of R is that the object returned by the log() function depends on the argument. Consider the following four lines of code:

```
x <- 2.72
log(x)
x <- c(4, 7)
log(x)
```

We can test the code in a new R script window. If you have worked through this section in a single session, then objects from the mark-recapture example will be in the Environment window. Start with a clean slate by clicking on the broom icon in the Environment window. As is often the case, this can also be accomplished through RStudio menu commands (Session : Clear Workspace) or R code rm(list=ls()). The first R statement assigns the value 2.72 to x, and the second statement prints the natural log of 2.72 in the Console window. The third line uses the c() (combine) function to create a vector of length two. Because we used the same variable name to store the result, R replaces the original integer value of x with a vector containing two integers. (It is done here as an example, but in the future, we will avoid this bad practice of reusing a variable name for a different purpose!) In the final line of code, the log() function returns a vector by applying the log() function element-by-element. This flexibility (allowing x to be defined first as a scalar then vector, and allowing either type as an argument to the log() function) is a strength of R, but it also increases the risk of coding errors compared to programming languages that are more structured. The keys to success in R programming are to check and test code carefully and to reuse existing code whenever possible. The chapters that follow will provide many more examples of R code, with comments and explanation as needed when new features are introduced. As you encounter R code, you should always run it to make sure that you understand how it works and its purpose.

---

[3]https://cran.r-project.org/doc/contrib/Short-refcard.pdf

3

Probability Distributions

When a biologist conducts a field study, the usual purpose is to collect data that will be useful for management. Examples of data that could be collected include (i) whether a fish with a radio transmitter was detected, (ii) how many tags were reported by anglers over a fishing season, (iii) how many fish of each age were observed in a sample, or (iv) number of fish caught in a single tow of a net. Each of these examples arises from a different statistical distribution. The distribution affects the properties of the data and how those data might be analyzed. For example, observations in the first example are either Yes or No, which could be represented as a 1 or 0. Having only those two possible values certainly affects characteristics of that distribution as well as methods for analysis.

In this chapter, we will examine some of the more common and important statistical distributions, and learn how to simulate observations from those distributions. Simulation will be an important skill in the chapters that follow, in that simulation provides a valuable way to evaluate field study designs and methods of analysis. The first few distributions that we consider are discrete (Section 3.1); that is, non-negative whole numbers that are appropriate for counts like the number of fish in a single haul of a net. The second category is continuous distributions (Section 3.2), for real numbers such as a catch *rate* of 0.52.

3.1   Discrete

3.1.1   Bernoulli

The most basic discrete distribution is the Bernoulli. There are only two outcomes, often referred to as success and failure. The classic example of a Bernoulli distribution is a coin toss, where "heads" might be considered success and "tails" failure. This distribution has a single parameter, p, representing the probability of success. Its complement (1-p) is the probability of failure. Searching a lake for a fish with a radio transmitter is a fisheries example of a

Bernoulli trial, where p represents the probability of detecting the fish's signal (success). Other biological examples are whether a fish with a transmitter is dead or alive and whether or not a study site is occupied.

Let's begin with R code for a hands-on example, where a coin was tossed ten times. It is convenient in R to use 0 for tails and 1 for heads:

```
y <- c(0,0,1,0,0,1,0,0,0,1)   # Coin toss observations, 0=tails
N.obs <- length(y) # R function to count number of observations
```

As usual, you should paste the code into an RStudio script window. You can select both lines and click the Run icon in order to run both at once. The first line assigns the vector of 0s and 1s to the variable y. The next line uses the length() function to count the number of observations in the vector y. We could have used the code N.obs <- 10, but there are two related reasons not to do so. First, hard-coding a constant for sample size means that we would need to remember to change that line if the y vector was later updated with a different number of observations. Second, you should never do anything manually that computer code can do for you. Letting R count the observations means that you will never miscount, and your code will still work correctly if you change the y vector and end up with a different sample size.

Another critical step in data analysis is to plot the data whenever possible, in order to look for patterns. Here, plotting the sequence of 0s and 1s is not that useful. A better summary plot is to examine the total number of 0s and 1s. We can use the table() function, which returns a vector with the number of 0s and 1s in the y vector (see Environment window). The barplot() function plots the two counts (Figure 3.1):

```
Counts <- table(y)   # Summary table of 0 vs 1
barplot(Counts)
```

Running these lines of code in a script window causes this plot to appear automatically under the Plots tab in the lower right RStudio window. We can save the plot image or copy and paste it into a report using menu option Export under the Plots tab. Examining the plot shows the unusual result of seven "tails" and three "heads". We would expect the counts to be about equal for the two categories, given the assumed probability of success (p) of 0.5 for a fair coin.

Can you think of a physical process (hands-on method) that would produce Bernoulli data using other values for p, such as 0.25 or 0.33? What value for p might you expect for a fisheries examples, such as an annual exploitation

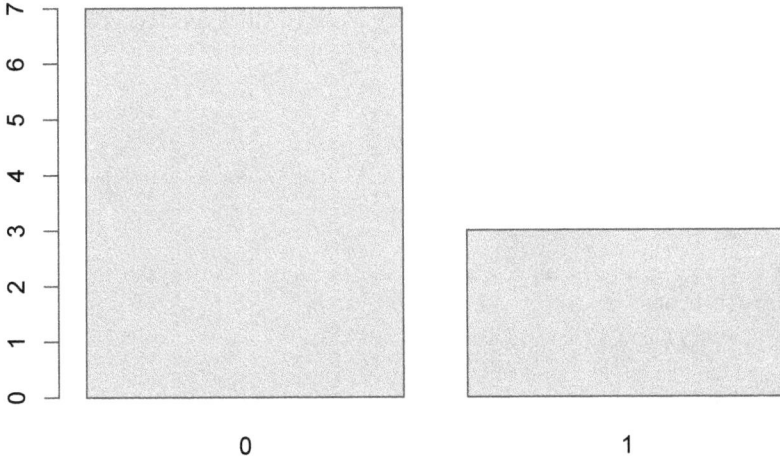

FIGURE 3.1 Summary barplot of coin tosses.

rate (probability of being harvested), tag-reporting rate (probability of angler returning a tag), or survival rate of adult fish in a lake (probability of surviving from year i to i+1)? These are the kinds of parameters that a field biologist might need to estimate. In the chapters that follow, we will examine some methods for collecting the data and estimating those probabilities.

We can generate simulated observations from a Bernoulli distribution using the `rbinom()` function:

```
N.obs <- 30
p <- 0.4
y <- rbinom(n=N.obs, prob=p, size=1)
Counts <- table(y)   # Summary table of 0 vs 1
barplot(Counts)
```

The `rbinom()` function generates values from a binomial distribution (described in Section 3.1.2). If we set the trial size to 1 (e.g., a single coin toss), we simulate draws from a Bernoulli distribution. The other two arguments in the `rbinom()` function are how many Bernoulli trials to simulate (n) and the probability of success (prob). How many values of 1 would you expect in a vector of 30 replicate observations? As above, the final two lines can be used to count 0s and 1s and view a plot for these simulated data. Try running the code several times to see the random variation in the total number of successes in 30 Bernoulli trials.

One of the most important benefits of using simulation is the ability to examine results from large sample sizes. For example, the following code shows that the

rbinom() function produces an outcome very close to the expected proportion
of 0s and 1s:

```
rm(list=ls()) # Clear Environment
# ls() function returns names of variables in Environment window

N.obs <- 1000
p <- 0.4
y <- rbinom(n=N.obs, prob=p, size=1)
SumOnes <- sum(y)
SumOnes
SumOnes/N.obs # Proportion of trials that are successful
Counts <- table(y)
barplot(Counts)
```

The first line of code uses the rm() (remove) function to clear the Environment,
ensuring that we start from a clean slate. The sum() function provides a total
for the y vector, which is equal to the number of successful trials. Dividing
by the number of observations is an empirical estimate of p. The plot (Plot
window) shows the result for one simulation (with N.obs replicates). How close
to the expected count of 1s (400) did you observe? Try running this block
of code several times. How does variation in the total number of successes
compare to the previous case with N.obs=30?

3.1.2   Binomial

The binomial distribution is simply the combined total from multiple Bernoulli
trials. One physical example would be the number of heads in five tosses of
a coin. Each toss is an independent Bernoulli trial, but we record the total
number of successes (heads). We could obtain anything between 0 and 5 heads
in a single trial. We assume that the probability of success (p) is constant for
each Bernoulli trial (e.g., each coin toss).

One fisheries example of a binomially-distributed process would be the number
of recaptures in a two-sample mark-recapture study. In that case, the number of
fish in the second sample is the size of the trial, and fraction of the population
that is marked is the probability p. Another fisheries example would be the
number of tag returns in a study to estimate the exploitation rate (rate of
harvest). The number of tagged fish is the size of the trial and the exploitation
rate is p. A telemetry study to estimate fish survival rate is a third example. The
initial number of fish with transmitters is the size of the trial. The survival rate
is p and the number of survivors at the start of the next period is binomially
distributed.

Let's start with a physical example of drawing five playing cards (with replacement) and recording the number of clubs. The probability of success in this case (drawing a club) is 0.25, and there are six possible outcomes (0-5 clubs). The expected number of clubs is 1.25 (trial size * p).

```
rm(list=ls()) # Clear Environment

Clubs <-c(0, 0, 3, 3, 1, 1)   # Drawing five cards, with replacement
N.trials <- length(Clubs) # Function for getting number of trials
Counts <- table(Clubs)
barplot(Counts)
barplot(Counts, xlab="Number of clubs")
```

Again we let R count the number of trials (6 in this case) and use the `table()` function to summarize the outcomes. Our sample size is small, and the plot only shows the three observed outcomes. The plot can be made more understandable (especially by someone else) by adding a plot option (xlab) to label the x-axis. Doing more than one plot within the RStudio environment produces a sequence of plots which you can view by using the blue left and right arrows in the Plot window. Practice modifying the above code by adding a label for the y-axis (e.g., "Frequency"). You can find the full list of barplot options by entering "barplot" in the search box in the RStudio help window (or simply type ?barplot in the Console).

We can alleviate our small sample size issue by turning to a simulation version. We again use the `rbinom()` function from the Bernoulli example, but now size is greater than 1:

```
rm(list=ls()) # Clear Environment

N.trials <- 30
p <- 0.25 # Probability of drawing a club
Trial.size <- 5
Clubs <- rbinom(n=N.trials, prob=p, size=Trial.size)
Clubs
Counts <- table(Clubs)
barplot(Counts)
```

Each of the 30 observations now represents a trial size of 5 (in total, a simulation equivalent to 150 Bernoulli trials). Only part of the Clubs vector is visible in the Environment window, but we can put Clubs as a line of code to print the whole vector in the Console. We again summarize the counts and plot the results using the `barplot()` function. Run this block of code multiple times to get some insight into the variability of the simulated system. For example, how often do you observe 0 or 5 Clubs? Which outcome occurs most often (mode) and how frequently does the mode change?

We would expect to observe very few trials with 5 successes, because the probability is quite low $(0.25^5=0.001)$. We can calculate it in the Console, by entering the expression 0.25^5. There is a higher probability of 0 successes (0.24), which can be calculated as 0.75^5 or $(1-p)^5$, because 1-p is the probability of failure (not getting a club). It is more involved to calculate the probabilities of 1-4 successes, because they can be obtained multiple ways (e.g., 1 success can be obtained five ways: 10000, 01000, 00100, 00010, 00001). We could calculate those probabilities using the binomial probability distribution: $f(k) = \binom{n}{k}p^k(1-p)^{n-k}$, where k is the number of successes in a trial of size n. R code for this calculation uses the `choose()` function; for example, `choose(Trial.size,0)` `* p^0 * (1-p)^5` calculates the probability of 0 clubs in five draws. We could get an approximate ("brute-force") estimate of the probabilities by increasing N.trials to a large value (say 10,000) and dividing the Counts vector by N.trials to estimate the proportion of trials resulting in 0, 1,... 5 successes:

```
rm(list=ls()) # Clear Environment

N.trials <- 10000
p <- 0.25 # Probability of drawing a club
Trial.size <- 5
Clubs <- rbinom(n=N.trials, prob=p, size=Trial.size)
#Clubs
Counts <- table(Clubs)
barplot(Counts)
Counts/N.trials # Vector division - probabilities of 0-5 Clubs
```

Note that we use a pound sign to turn off ("comment out") the line for printing the Clubs vector to the Console, and print to the Console the estimated probabilities. Dividing the Counts vector by a constant produces a vector of estimated probabilities. The mode of our simulated distribution is in agreement with the expected value (trial size * p), as are the estimated probabilities for 0 and 5 clubs.

3.1.3   Multinomial

The Bernoulli and binomial distributions return a single value: success or failure for the Bernoulli and the number of successes (out of a specified trial size) for the binomial. The multinomial distribution extends that to a vector of results. A classic physical example would be to roll a die several times and record how many rolls were a 1, 2, ..., 6. A fisheries example would be to collect a sample of fish and determine the number of fish by age. The focus in that example is on estimating the age distribution, or proportions of fish by age. An age distribution contains valuable information about the rates of survival and recruitment (year-class strength). Another fisheries example would be to collect a sample of fish (e.g., from electrofishing or a trawl) and determine the number caught by species. Species composition can be an important indicator of the ecological health of a study site (e.g., if there is a high relative abundance of species considered tolerant of poor water quality).

Let's begin with a hands-on example. I roll a die ten times, and record as a vector the number of times I get a 1, 2, ..., 6:

```
rm(list=ls()) # Clear Environment

x <-c(2,3,3,0,2,0)    # Generated using die, equal cell probabilities
SampleSize <- sum(x)
barplot(x)
```

Here R calculates the sample size (SampleSize), even though we know in this instance that there were ten rolls. The `barplot()` function generates the plot, but without any labels it is not very appealing. We can dress it up a bit by adding some options (Figure 3.2). We first create a vector with values 1-6 to serve as x-axis labels (`seq()` function, from 1 to 6 by 1). The barplot() function uses the names.arg option to add the labels, with axis titles added using xlab and ylab. The final option yaxt="n" suppresses the y-axis. Normally the default plotting options are fine, but adding the y-axis as a separate step with the axis() function allows us to use integer labels (given that the number of rolls will only be a whole number). The axis() arguments specify that the axis is added to the left side of the plot (side=2) and provide the vector of tick-mark positions. One general note about plotting is that only base R plots are used in this book. They are sufficient for our purposes, but fancier plots can be produced using graphical packages such as **lattice** and **ggplot2**.

```
x.labels <- seq(from=1, to=6, by=1) # x-axis labels for plotting
barplot(x, names.arg=x.labels, xlab="Die side", ylab="Number of rolls",
        yaxt="n")
axis(side=2, at=seq(from=0, to=3, by=1))
```

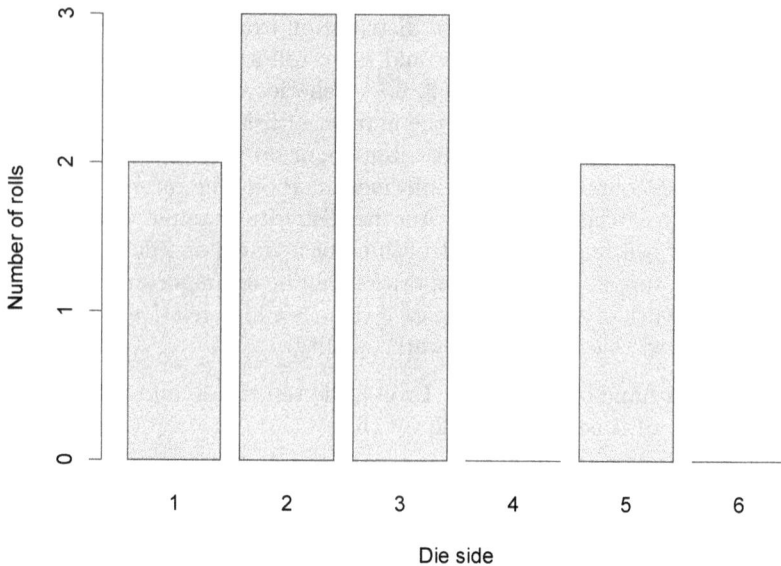

FIGURE 3.2 Summary bar plot of die rolls.

We can simulate a multinomial distribution as follows:

```
rm(list=ls()) # Clear Environment

SampleSize <- 30   # e.g., sample of 30 fish assigned by age
TrueP <- c(0.1, 0.3, 0.4, 0.1, 0.05, 0.05)
   # probability of being ages 1-6
Reps <- 1 # How many replicate multinomial trials to carry out

# Simulate a single trial using rmultinom function
rmultinom(n=Reps, prob=TrueP, size=SampleSize)
   # Prints matrix to Console
x <- rmultinom(n=Reps, prob=TrueP, size=SampleSize)
# x contains matrix of counts (one column per Rep)
x.vec <- as.vector(t(x)) # Transpose matrix then convert to vector

age.vec <- seq(1,6, by=1)
barplot(x.vec, names.arg=age.vec, xlab="Age", ylab="Number of fish")
```

In this simulation, the age proportions (TrueP) are arbitrarily chosen (but could be based on field data). Run statements one-by-one to see how each line of code works. Note that the rmultinom() function returns a matrix with six rows and one column (n determines how many columns are returned).

Rather than a column, we want a row vector, with the number of fish by age in successive columns. We use the t() (transpose) function to transpose the matrix and the as.vector() function to convert the matrix (with one row) into a row vector. We could have done this transformation in two steps but instead have nested the two functions. That results in more compact code, but it is always a fine option to do the steps separately for greater clarity.

Run the code beginning with the x <- rmultinom() line several times to see how much the age distribution varies from one random sample to the next. The underlying age proportions are fixed, but the counts vary quite a bit for this small sample size. This is an example of how simulation is helpful in gaining experience and intuition in judging pure sampling variation versus a real biological result (e.g., good or bad year-class strength).

3.1.4 Poisson

Like the Bernoulli, binomial, and multinomial distributions, the Poisson and negative binomial distributions are used for counts (non-negative whole numbers). The first three cases have an upper bound (1 for the Bernoulli, trial size for the next two), whereas the Poisson and negative binomial distributions do not. The Poisson is simpler than the negative binomial in that it is described by a single parameter, usually λ, which represents the mean and variance of the distribution. Fisheries examples of a Poisson distribution could be the number of fish per hour moving upstream via a fish ladder, or the number caught per trawl tow.

Our simulation uses an arbitrary value for λ. The rpois() function has two arguments: sample size (n) and lambda, representing the mean and variance of the distribution.

```
rm(list=ls()) # Clear Environment

N <- 30
lambda <- 4
Count <- rpois(n=N, lambda=lambda)
Freq <- table(Count)   # Distribution of simulated counts
barplot(Freq, main="", xlab="Count", ylab="Frequency")

mean(Count)
var(Count)
```

Run the lines of code multiple times to gain experience with the Poisson distribution. How often is the mode of the distribution equal to λ? Are the estimated mean and variance (printed to the Console) close to λ? Increase the sample size to get a smooth distribution and reliable estimates of the two sample statistics.

It is also useful to run the code using other values for λ. What value for λ causes the mode to shift to 0? At what value does the sample distribution look like a bell curve (i.e., symmetrical)? One of the key benefits of simulation is that we can easily try different scenarios to build understanding about the system being modeled.

Especially for smaller sample sizes, the plot can be misleading because it only includes observed levels. We can modify the code to include levels with a frequency of zero:

```
rm(list=ls()) # Clear Environment

N <- 30
lambda <- 4
Count <- rpois(n=N, lambda=lambda)
Freq <- table(factor(Count, levels = 0:max(Count)))
barplot(Freq, main="", xlab="Count", ylab="Frequency")

mean(Count)
var(Count)
```

Now the `table()` function operates on counts transformed into factors (categories or levels) by the `factor()` function. The levels= argument forces factor() to use a range of levels from 0 to max(Count), in order to include levels with a frequency of zero. Try the following partial code in the Console to see the levels ("0", "1", etc.) created by the `factor()` function: factor(Count, levels = 0:max(Count)).

3.1.5   Negative binomial

As noted above, the negative binomial distribution is used in similar situations to the Poisson, except that it has two parameters and is therefore more flexible. This is particularly helpful for ecological data, where counts are often more variable than would be expected under a Poisson distribution (Bolker, 2008; Link and Barker, 2010). The R function for generating negative binomial random variates (`rnbinom()`) can be parameterized in different ways; here I use the "ecological" parameterization recommended by Bolker (2008). This uses the mean (termed μ) but not the variance, instead using an overdispersion parameter k. Lower values of k result in a more heterogeneous distribution, and in ecological settings, k is often less than the mean (Bolker, 2008). The distribution approximates a Poisson when k is large (e.g., $>10*\mu$). The code includes calculation of the variance as a function of mu and k (Bolker, 2008). Alternatively, the mean and variance (V) can be specified and used to solve for k: $k = \mu^2/(V - \mu)$. Note that overdispersion implies that the variance is greater than the mean, so that the denominator is greater than zero.

```
rm(list=ls()) # Clear Environment

N <- 30
mu <- 4
k=1
variance <- mu+(mu^2)/k
Count <- rnbinom(n=N, mu=mu, size=k)
Freq <- table(factor(Count, levels = 0:max(Count)))
barplot(Freq, main="", xlab="Count", ylab="Frequency")

mean(Count)
var(Count)
```

This distribution has the same mean as in the Poisson example but tends to have a longer tail (occasional large counts). Rerun the code using different values for k; for example, Figure 3.3 uses a sample size of 200 to compare results for k=1 and k=10, each with μ=4.

3.2 Continuous

3.2.1 Uniform

A uniform distribution is flat in shape, indicating that all values between a lower and upper bound are considered equally likely. It is also useful as a "first" model for situations where relatively little is known other than the lower and upper bounds (Law and Kelton, 1982). Probabilities have default bounds of 0 and 1, so a fish population model might use a uniform 0-1 distribution for probabilities such as survival rate. More narrow bounds could be used in cases where additional information was available; for example, a pilot study might indicate that the tag-reporting rate would be expected to vary between 0.4 and 0.6. Another fisheries example would be a uniform distribution between 60 and 70, for simulating individual variation in maximum size for a growth curve.

We begin with a simulation example, using the default range of 0-1 for the `runif()` function:

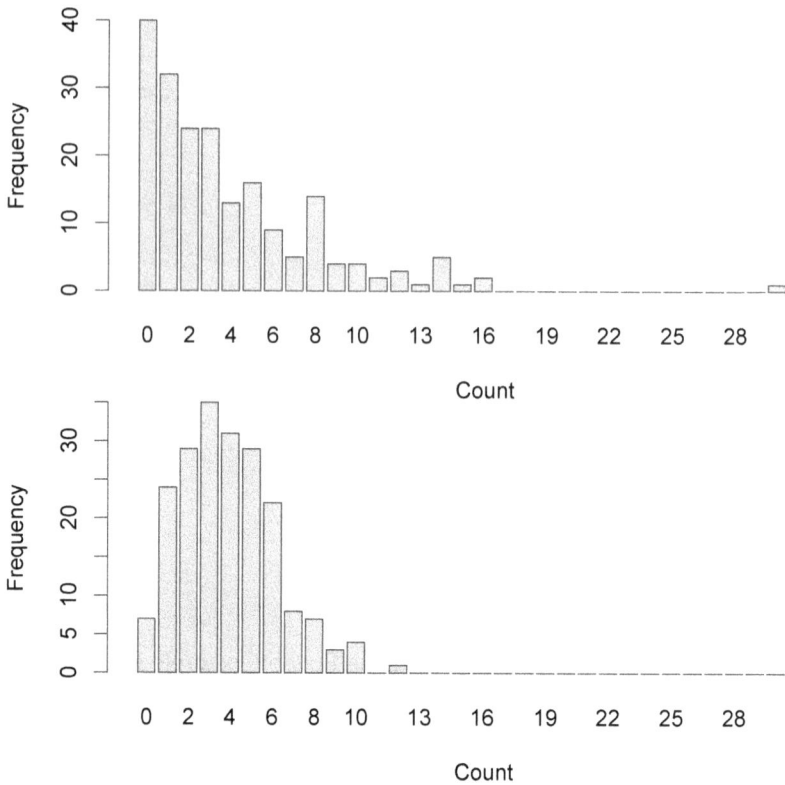

FIGURE 3.3 Negative binomial frequency distributions for mu=4 and k=1 (top) or 10 (bottom).

```
rm(list=ls()) # Clear Environment

Reps <- 30 # Replicate draws from a uniform distribution
p <- runif(n=Reps)# Default range of 0-1. Can specify using min, max
hist(p, breaks=5, main="")
```

Try running the code several times and note the variation in pattern from one simulation run to the next. I use the breaks=5 option in the hist() function, so that the histogram bin intervals are stable among runs. Specifying a null character string for "main=" prevents the plot from including a main title. Next, increase the sample size and note the increased smoothness of the observed distribution. What would be the expected mean of the distribution? Estimate the mean and compare the observed and expected value among runs.

As noted above, the runif() function allows for a uniform distribution using any lower and upper bounds. For example, we might simulate uncertainty about an exploitation rate (fraction of the population that is harvested) by allowing randomly drawn values in a specified interval, for example, `runif(n=Reps, min=0.3, max=0.4)`. The `runif()` function is not constrained to positive values; for example, it could be used with bounds -0.5 and 0.5 to simulate uncertainty about a parameter that is close to 0 but can be positive or negative. The chapters that follow will include many examples where a uniform distribution is relevant for fisheries modeling.

3.2.2 Beta

The beta distribution has a lower bound of 0 and an upper bound of 1, so it is useful for modeling parameters with those bounds (i.e., probabilities). A fisheries model might use a beta distribution to simulate variation in a survival rate or the probability of a tag being returned. The uniform distribution (Section 3.2.1) can also be used for probabilities, but only for the case where all values are equally likely. The beta distribution is more flexible and can not only be flat but also u-shaped, bell-shaped, or left- or right-skewed. Figure 3.4 shows sample distributions for shape parameters $\alpha=1$ and $\beta-0.5$, 1, and 5.

We simulate draws from a beta distribution using the `rbeta()` function. Starting with α and β set to 1 would be appropriate for a situation where all probabilities are equally likely:

```
rm(list=ls()) # Clear Environment

Reps <- 200 # Replicate draws from a beta distribution
alpha <- 1
beta <- 1
x <- rbeta(n=Reps, shape1=alpha, shape2=beta)
hist(x, main="", xlab="Probability")
```

After running the above code multiple times, try a beta(8, 4) distribution. These parameter values could result from a pilot study with seven successes in ten trials ($\alpha=7+1$, $\beta=3+1$). This shifts the distribution toward a mode of 0.7, which is the observed proportion of successes in ten trials. Try other arbitrary values for the two parameters to vary the shape of the distribution. Can you think of a fisheries situation where probabilities would be expected to be close to 0 or 1?

3.2.3 Normal

The normal distribution is the well-known bell curve from traditional statistics courses. The distribution is unbounded (bounds $-\infty$, ∞). The shape of the

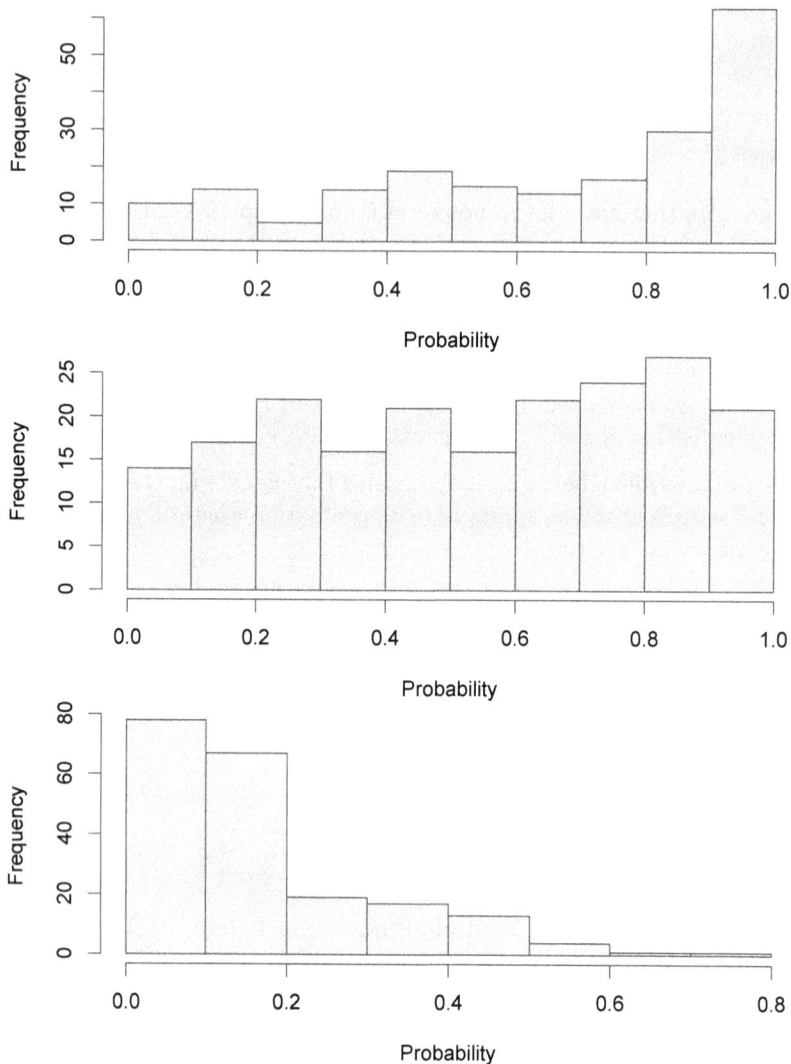

FIGURE 3.4 Histograms for 200 random draws from a beta distribution, with $\alpha=1$ and $\beta=0.5$ (top), 1 (middle), and 5 (bottom).

distribution is determined by the mean (location) and standard deviation (spread). A fisheries example might be a simulation where size-at-age varied among individuals according to a normal distribution.

Let's look at simulation code for length of individual fish, assuming a mean of 30 and standard deviation of 2:

```
rm(list=ls()) # Clear Environment

Reps <- 30 # Replicate draws from a normal distribution
Len <- rnorm(n=Reps, mean=30, sd=2)
hist(Len, main="")
```

Run the code a few times to look at the variation in shape and bounds at a relatively small sample size. Next, increase the sample size and determine through trial and error the smallest sample size that provides a relatively smooth bell-curve shape. Estimate the mean and standard deviation (function `sd()`) and compare the observed and true values.

3.2.4 Lognormal

The lognormal distribution (bounds $0, \infty$) is useful in ecological settings, because it excludes negative values and allows for a long upper tail. A classic fisheries example of a lognormal distribution is annual recruitment for marine fish species (Hilborn and Walters, 1992). Most years result in low recruitment, but there are occasional extreme values when environmental conditions are right.

We begin with simulation code:

```
rm(list=ls()) # Clear Environment

Reps <- 200
ln.m <- 0 # Ln-scale mean
ln.v <- 0.5 # Ln-scale variance
y <- rlnorm(n=Reps, mean=ln.m, sd=sqrt(ln.v))

hist(y, main="Lognormal distribution")
Exp.mean <- exp(ln.m+(ln.v/2)) # Arithmetic scale expected mean
abline(v=Exp.mean, lty=3, lwd=3)
Exp.var <- (exp(2*ln.m+ln.v))*(exp(ln.v)-1)
   # Arithmetic scale expected variance
mean(y)
var(y)
```

The `rlnorm()` function generates random lognormal values, using an arbitrarily chosen mean and standard deviation specified in natural log-scale. Those values can be used to calculate the expected mean and variance in arithmetic scale. The `abline()` function adds to the histogram a vertical (v=) line for the expected arithmetic-scale mean (Figure 3.5). The line type is specified as dotted (lty=3) and weight 3 (lwd=3); other options including line color can be found in the Help window. Compare the estimates of arithmetic-scale mean and variance for your (relatively large) sample of lognormal random variates to the expected values (visible in the Environment window).

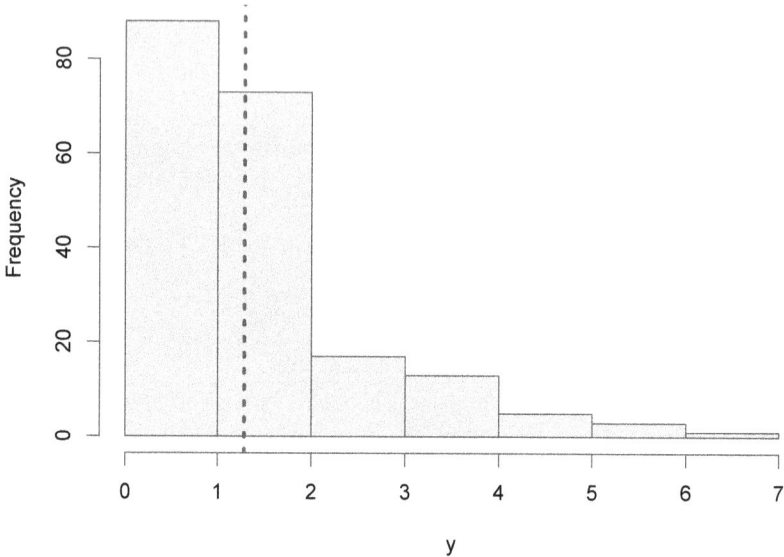

FIGURE 3.5 Histogram for 200 lognormally distributed values, with ln-scale mean of 0 and variance of 0.5. Dotted vertical line denotes expected mean.

3.2.5   Gamma

The gamma distribution is similar to the lognormal in that it has a lower bound of 0 and an upper bound of ∞. Thus it also excludes negative values and allows for a long upper tail. This distribution is often used to simulate waiting times (how long until some number of events has occurred, based on a specified average time between events). We consider it here simply because it can take on a variety of shapes depending on its parameters (Bolker, 2008). Figure 3.6 shows sample distributions for shape parameter $\alpha=3$ and scale or spread parameter $\beta=1$ and 3.

We simulate draws from a gamma distribution using the `rgamma()` function, with arbitrary values for α and β:

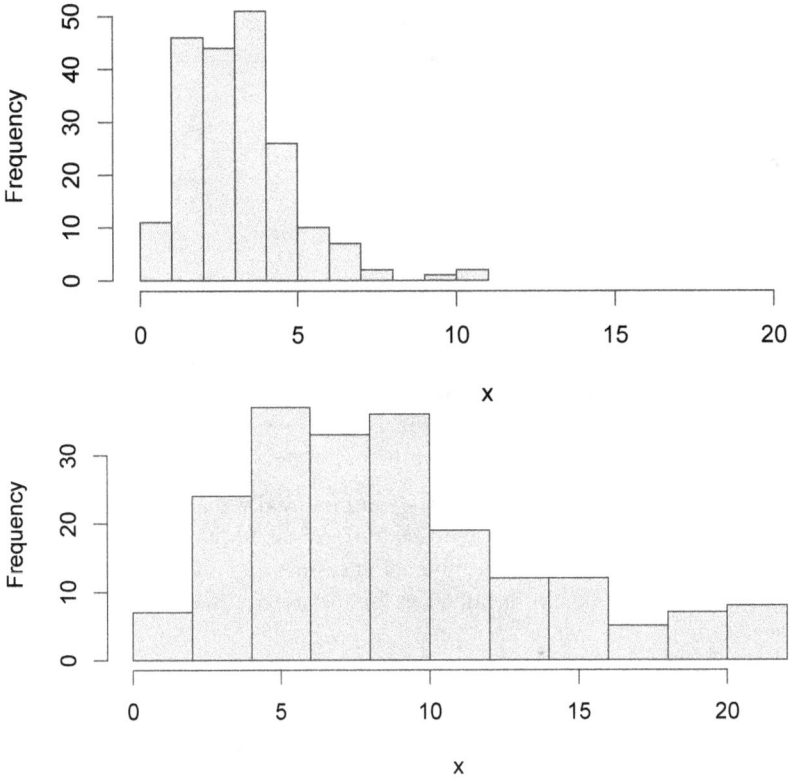

FIGURE 3.6 Histogram for 200 gamma-distributed values, with $\alpha=3$ and $\beta=1$ (top) and 3 (bottom).

```
rm(list=ls()) # Clear Environment

Reps <- 200 # Replicate draws from a gamma distribution
alpha <- 3 # Shape parameter, > 0
beta <- 3 # Scale parameter, > 0
x <- rgamma(n=Reps, shape=alpha, scale=beta)
hist(x, main="")
Exp.mean <- alpha*beta
Exp.var <- alpha*beta^2
mean(x)
var(x)
```

A sample size of 200 replicates produces a relatively smooth pattern for the distribution. How close are the estimated mean and variance to the expected values for this sample size? Try adjusting the two parameters to vary the shape of the distribution. What values for α and β produce a right-skewed distribution (long right tail), or a pattern somewhat like a normal distribution (bell-curve)?

3.3 Exercises

1. Use a physical process (i.e., not simulated) to generate 20 Bernoulli observations with probability of success 0.25. Include in your code comments describing how you obtained the data. Produce a summary plot of the total number of 0s and 1s. How does your observed number of successes compare to what you would expect?

2. Use a physical process to generate ten binomially distributed observations, with a trial size of at least four. What is the probability of success and the expected number of successes per trial? Produce a histogram showing the frequencies for different numbers of success per trial.

3. Use a physical process to generate ten multinomial trials with at least three possible outcomes. Describe in your code comments the physical process for obtaining the data and include a plot of the frequencies. How do the frequencies compare to what you would expect?

4. Assume that the following vector contains trawl catches: y=c(7, 5, 1, 5, 0, 3, 8, 2, 14, 5, 0, 6, 9, 1, 2, 2, 1, 1, 0, 9, 2, 2, 4, 11, 1, 1, 7, 3, 1, 1). Based on your analysis (including plot), were these data likely generated from a Poisson or negative binomial distribution?

5. Provide simulation code for generating normally distributed length data for fish ages 1 (n.1=40, age-1 mean 30, sd 5) and 2 (n.2=25, age-2 mean=50, sd=7). Plot the combined length sample. In multiple simulations using these default values, how frequently is the length distribution visibly bimodal?

4

Model Fitting

4.1 Introduction

What is a model in a fisheries context? It typically means one or more equations that describe a biological process of interest. There are several good reasons for developing a model. A model that simulates a field study can be used for planning purposes, before field work is conducted. Such a model might be used to determine the number of fish to tag or the level of electrofishing effort, in order for the field work to produce a useful result. A model can also be used to test an analytical approach. For example, we could generate simulated data and see whether our analysis agrees with the known true value(s). The other main use of a model is to analyze data from field work. One example from Chapter 2 is a two-sample mark-recapture study to estimate population size. Fish in the first sample are tagged, and the fraction of marked fish in the second sample provides the information needed to estimate population size. The model in that case is the equation for population size (N_{hat}), along with some assumptions about sampling. The equation is based on the assumption that the marked fraction of the second sample should (on average) be the same as the marked fraction in the underlying population. This method also requires some other assumptions that are more practical in nature, to help ensure that our field methods approximate the underlying model structure. For example, we assume that tagged and untagged fish are well mixed. In practice, this means that tagging effort should be well distributed throughout the study area so that our recapture effort does not encounter pockets with too many or too few tagged fish. We also assume that tags are not shed and that there are no deaths due to tagging. The mark-recapture model assumes that we know how many fish are in the tagged subpopulation. Tag loss or tagging mortality affect that number and would introduce bias.

A key result of model fitting is estimating the level of uncertainty in parameter estimates. The sections that follow illustrate two approaches for estimating that uncertainty, first through simulation then a Bayesian statistical framework. Hopefully the comparison between approaches will be helpful in understanding how a Bayesian analysis works.

4.2 Using simulation

The equation for a two-sample mark-recapture study provides a point estimate of population size, but we could improve the model by including the sampling process. For example, we could assume that the number of marked fish in the second sample comes from a binomial distribution. That distributional assumption allows us to generate estimates of uncertainty as well as the point estimate. The key to this extended model is to think about the process generating the field data and determine which statistical distribution is appropriate.

We begin with the deterministic version, using the same estimator as in Chapter 2. The two samples could be fixed values and could differ, but here we assume they are equal in size, based on the capture probability (p.true). The number of recaptures in the second sample equals the expected number based on the fraction of the population that is marked (p.true). Because m$_2$ takes on its expected value based on p.true, the estimate (N$_{hat}$) equals the true value (see Environment window).

```
rm(list=ls()) # Clear Environment

# Two-sample population estimate
N.true <- 400   # Arbitrary assumed population size
p.true <- 0.2 # Capture probability (fraction of population sampled)
n1 <- N.true * p.true # Caught and marked in first sample
n2 <- N.true * p.true # Caught and examined for marks in second sample
m2 <- n2 * p.true # Marked fish in second sample
N.hat <- n1*n2/m2 # Estimated population size
```

Now, let's extend the model by introducing random variation (stochasticity) in the number of recaptures. We assume that recaptures are the successes in a binomial trial and replace the deterministic calculation for m$_2$ with the following line of code, using the rbinom() function:

```
m2 <- rbinom(n=1, prob=n1/N.true, size=n2) # Marked fish, second sample
```

The size of the binomial trial is n$_2$. The probability of success equals p.true here, but we calculate it as n$_1$/N.true in case n$_1$ and n$_2$ were given other fixed values. Running the block of code simulates a single mark-recapture experiment, with N.hat varying randomly according to the binomially distributed sampling process.

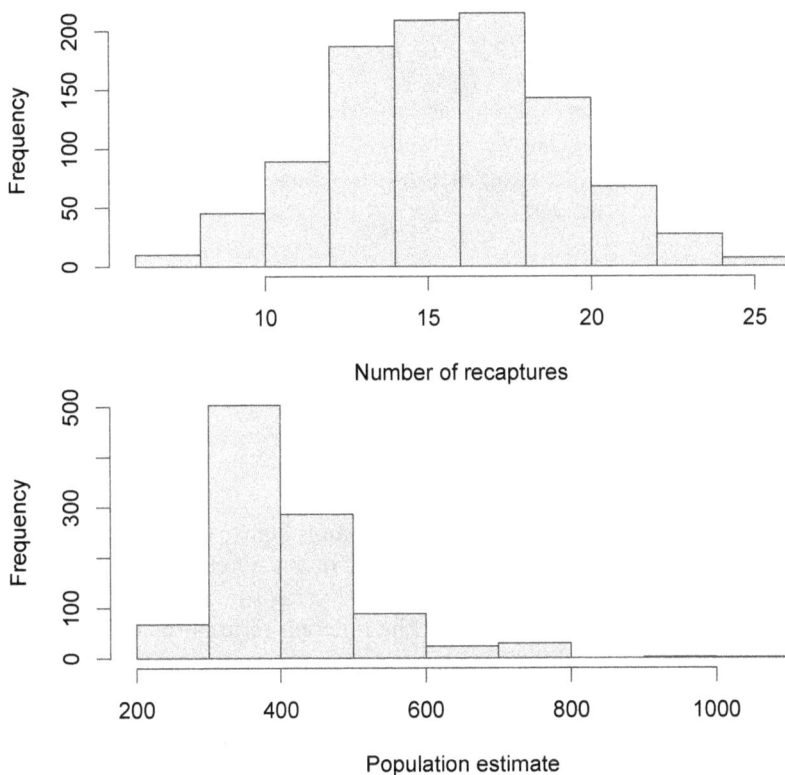

FIGURE 4.1 Frequency distributions for number of recaptured fish (top) and estimated population size (bottom), based on a simulated mark-recapture study using a binomial distribution for the number of recaptures.

An easier way of judging the reliability of this mark-recapture study is to view results from a large number of replicate experiments. Using n=Reps in the rbinom() function produces a vector of recaptures, simulating a mark-recapture study repeated Reps times. Using that vector in the equation for Nhat.vec now produces a vector of population estimates. Histogram plots (Figure 4.1) shows that estimates are generally close to the true value but tend to have a long tail to the right (overestimating N.true when a low value is drawn for the number of recaptures).

```
rm(list=ls()) # Clear Environment

# Extended model to include stochasticity
N.true <- 400   # Arbitrary assumed population size
p.true <- 0.2 # Capture probability (fraction of population sampled)
```

```
n1 <- N.true * p.true # Caught and marked in first sample
n2 <- N.true * p.true # Caught and examined for marks in second sample

Reps <- 1000
m2.vec <- rbinom(n=Reps, prob=n1/N.true, size=n2)
Nhat.vec <- n1 * n2 / m2.vec

hist(m2.vec, xlab="Number of recaptures", main="")
hist(Nhat.vec, xlab="Population estimate", main="")

mean(Nhat.vec)
quantile(Nhat.vec, probs=c(0.025, 0.5, 0.975), na.rm=TRUE)
# quantile function provides empirical confidence bounds and median
```

Results printed to the Console show that the mean is higher than the true value (due to the tail of extreme values), but the median is very close to N.true. The estimated median and 95% bounds are obtained using the quantile() function. We provide a vector of probabilities, and the function returns the estimates at those points. The 0.025 and 0.975 points provide the empirical 95% bounds, and the 0.5 point is the median.

4.3   Using Bayesian methods

In Section 4.2, we used a simulation-based or numerical approach to estimate confidence limits for the mark-recapture study design. This is an example of parametric bootstrapping (Efron and Tibshirani, 1993), where replicate samples (number of recaptured fish in this case) are drawn from a parametric distribution, in this case binomial (Section 3.1.2). We can compare this simulation approach to a formal statistical method, using a Bayesian framework. We use the same binomial distribution to describe the process of obtaining recaptures. Our results will be similar, and hopefully the comparison will provide some intuition about how a Bayesian analysis works. Our use of simulation in the remainder of the book will be limited to generating "sample" data. Model fitting will be done using Bayesian methods that work well for simple or complex models.

The two main paradigms for statistical methods are frequentist and Bayesian. Frequentist methods are the classical approach for estimating model parameters and making inferences (Link et al., 2002; Royle and Dorazio, 2008). In a frequentist framework, model parameters are assumed to be fixed but unknown values. For example, we could conduct a two-sample mark-recapture study to estimate population size in a lake. If we repeated the mark-recapture study many times, we would expect the 95% confidence limit for each estimate to contain the true population size 95% of the time (hence the term "frequentist").

A Bayesian analysis is based on the data in hand and does not rely on the idea of hypothetical replicate studies (McCarthy, 2007; Royle and Dorazio, 2008). The results are exact for any sample size (Kéry and Schaub, 2012), which is an important advantage for fisheries studies because sample sizes are often small. Bayesian statistics also differs from frequentist statistics in that it takes into account what is known before the study is conducted (McCarthy, 2007). The requirement to specify prior knowledge can be viewed as either an advantage or disadvantage, depending on the availability of prior data and one's statistical philosophy (Ellison, 2004; McCarthy, 2007; Kéry, 2010; Kéry and Schaub, 2012; Dorazio, 2016; Doll and Jacquemin, 2018; Banner et al., 2020). The prior information takes the form of a statistical distribution that defines the range and relative likelihood of possible values, prior to collecting the new data. The prior distribution, or simply prior, can range from uninformative to informative. For example, a uniform 0-1 distribution would be a uninformative prior for a tag-reporting rate, because probabilities are by definition between 0 and 1. Results from a pilot study might allow for an informative prior, for example, a uniform distribution between 0.2 and 0.4.

Bayes' rule provides the formal structure for combining the prior information and new data in order to estimate the posterior distribution (Link et al., 2002; Kéry and Schaub, 2012). For our purposes, it is sufficient to know that the posterior distribution is proportional to the product of the likelihood (based on the new data; Section 4.3.4) and the prior:

$$posterior \propto likelihood \cdot prior$$

(Lunn et al., 2012). The details in applying Bayes' rule are handled automatically by the software used in this book. Nevertheless, knowing this relationship helps in understanding how the two components work together. Being required to specify the prior distribution makes a Bayesian analysis very transparent (Kéry and Schaub, 2012). Anyone reviewing the results can decide about the appropriateness of the prior(s). This would be particularly important in cases where a prior was based on belief (e.g., expert opinion) rather than prior studies.

For most analyses in this book, I adopt the common approach of using flat or vague priors that are intended to be uninformative (Royle and Dorazio, 2008; Kéry and Schaub, 2012). This approach is sometimes referred to as an objective Bayes analysis (Link et al., 2002). The simplest example is using a uniform(0,1) distribution for probabilities. In cases where choosing an uninformative prior distribution is more subjective (e.g., average maximum size for a growth curve), the analysis can be repeated using narrower and wider priors to ensure that results are insensitive to the bounds (Kéry and Schaub, 2012). The ideal situation is for the likelihood term to dominate, which is often referred to as the "data overwhelming the prior." This would mean that the sample size for new data was sufficiently large that the prior had little or no influence on results. Typically the results obtained using uninformative priors will be very similar to results of a frequentist analysis (Link et al., 2002; McCarthy, 2007; Kéry, 2010; Kéry and Schaub, 2012).

Banner et al. (2020) encourage investigators to think carefully about what is known and to take advantage of the Bayesian capability of using prior information, rather than automatically using "default" priors such as a uniform 0-1 distribution for a probability. Informative priors can improve the precision (and sometimes accuracy) of study results (e.g., Doll and Jacquemin, 2018). Considerable judgment may be required when selecting relevant data for a prior (Lunn et al., 2012). A fisheries example could be whether to use growth information from a prior year when population size was higher. Another case could be whether to use a tag-reporting rate from a prior year, if over time there was increased angler dissatisfaction with the current management approach. Sections 7.1.1 and 8.1.1 provide additional information about choosing a prior distribution and the use of an informative prior.

4.3.1   Example: mark-recapture

The following code provides for a Bayesian model fit to the mark-recapture data, using R packages for JAGS (Plummer, 2003). The code should transfer with little or no modification to other BUGS-family software versions.

```
rm(list=ls()) # Clear Environment

# Arbitrary 'observed' values for analysis
N.true <- 400   # Population size
p.true <- 0.2 # Capture probability (fraction of population sampled)
n1 <- N.true * p.true # Caught and marked in first sample
n2 <- N.true * p.true # Caught and examined for marks in second sample
m2 <- n2 * p.true # Marked fish in second sample

# Load necessary library packages
library(rjags)   # Package for fitting JAGS models from within R
library(R2jags)  # Package for fitting JAGS models. Requires rjags

# JAGS code
sink("TwoSampleCR.txt")
cat("
    model {
```

```
    # Prior
    N.hat ~ dlnorm(0, 1E-6) # uninformative prior (N.hat>0)
    MarkedFraction <- n1/N.hat

  # Likelihood
    # Binomial distribution for observed recaptures
    m2 ~ dbin(MarkedFraction, n2)
    }

    ",fill = TRUE)
sink()

# Bundle data
jags.data <- list("n1", "n2", "m2")

# Initial values.
jags.inits <- function(){ list(N.hat= runif(n=1,
                                  min=max(n1,n2), max=2000))}

model.file <- 'TwoSampleCR.txt'

# Parameters monitored
jags.params <- c("N.hat", "MarkedFraction")

# Call JAGS from R
jagsfit <- jags(data=jags.data, inits=jags.inits, jags.params,
                n.chains = 3, n.thin = 1, n.iter = 2000,
                n.burnin = 1000, model.file)
print(jagsfit, digits=3)
plot(jagsfit)
```

The first section of code sets up the experiment, with arbitrary values for population size and capture probability (p.true). The "observed" values for n_1, n_2 and m_2 could be replaced by real data from a field study; in that case, N would be unknown.

Before the first time of running JAGS code, you will need to download and install the current version of JAGS[1]. Next, use RStudio to install the packages for running JAGS. There are several options; this book uses **rjags** (Plummer, 2024) and **R2jags** (Su and Yajima, 2024). Both of these steps are done only once. Then each time that you want to access JAGS from R, you load the two packages using the library() function.

---

[1] https://sourceforge.net/projects/mcmc-jags/files/

The cat() function concatenates (merges together) the lines of JAGS code, which are written out as a text file using the sink() function. The name of the external file is arbitrary but just needs to match the model.file code further below. It can be helpful to use an informative name for the text file rather than something generic that might get overwritten or be hard to locate at a later date. The JAGS code can be thought of as consisting of two main parts: prior information and analysis of the new data. These two parts can come in any order, but it seems logical to put the prior information first. In this case, our model parameter, N.hat, can only take on positive values so we use an uninformative lognormal (McCarthy, 2007) as the prior distribution. The analysis of new data uses the likelihood (see Section 4.3.4) which is calculated in the final line of JAGS code. The observed number of recaptures is assumed to be binomally distributed. The parameter N.hat is used to estimate the fraction of the population that is marked, which is the probability for the binomial function dbin().

The next few lines of R code provide settings for the JAGS analysis (data to pass in, initial values for parameters, file name for the JAGS code, parameters to return). Using a lognormal distribution to generate initial values for N.hat occasionally led to convergence issues so initial values were obtained from a more restricted uniform distribution. The lower bound can be set using the max() function and the observed sample sizes, and the upper bound can be based on biological knowledge. Parameters (true model parameters and functions of those parameters) for which we want JAGS to monitor and return results are listed in jags.params. The function call to jags() links to the data, initial values, etc., and provides settings for the estimation process.

JAGS uses simulation techniques referred to as Markov Chain Monte Carlo (MCMC) methods to draw samples of each parameter (McCarthy, 2007; Kéry and Schaub, 2012). Each sample in a Markov Chain depends on the previous sample, so successive values are autocorrelated. Given a sufficient number of draws, the sampled distribution approximates the posterior distribution for that parameter. To judge convergence to the posterior distribution, the usual practice is to use multiple chains from separate streams of pseudo-random numbers and starting from different initial values. Three chains is a good practical choice. Thinning (e.g., n.thin=10 to retain only every tenth MCMC result) is sometimes done to reduce the autocorrelation or memory requirements for long runs (Gelman and Hill, 2007), but it is not generally necessary (Link and Eaton, 2012).

The number of MCMC iterations is arbitrary, but it is better to err on the side of having too many (and can provide a valuable opportunity to stretch legs or get coffee). R returns a measure of convergence (Rhat) that can be useful in deciding if a larger number of iterations is needed. The estimate of Rhat is somewhat like an analysis of variance, in that it is a ratio comparing between- and within-chain variability (Kéry and Schaub, 2012). The two types of variability should be similar if chains have converged and are providing samples from the posterior distribution. Rhat values less than 1.05 (Lunn et al., 2012) or 1.1 (Gelman and Hill, 2007) can indicate acceptable convergence, with values closer to 1 being preferred. More details about judging convergence are in Section 4.3.3. The final MCMC setting is for the "burn-in" phase, where initial estimates are discarded, so that the retained estimates are considered to be unaffected by the initial values. The default if n.burnin is not specified is to discard half the total number of updates. The best way to develop experience and intuition about MCMC settings (and Bayesian analysis in general) is to run and rerun code using different settings. For example, are the results different if n.iter is set to 10000?

Results for this case are contained in the arbitrarily named 'jagsfit' object, which can be printed to the Console and viewed as plots. Summary statistics are produced for the two monitored parameters and deviance (Section 4.3.5), which is a measure of fit. Convergence was very good based on estimated Rhat values less than 1.01. The printed output includes n.eff, which is an estimate of the effective sample size. The effective sample size is an estimate of what the sample size would be if the simulation draws were independent (rather than autocorrelated MCMC values). Gelman and Hill (2007) recommend an effective sample size of at least 100.

Summary statistics include the 2.5 and 97.5% percentiles from the estimated posterior distribution, which I use as a 95% Bayesian credible interval. This range has a 0.95 probability of containing the true value (McCarthy, 2007). Bayesian credible intervals and frequentist confidence intervals are typically very similar if Bayesian priors are uninformative (McCarthy, 2007; Kéry and Schaub, 2012). The credible interval in one simulation of our mark-recapture study was 277 to 670, with a median of 410 and mean of 425. Your results will differ slightly because of the pseudo-randomness of the MCMC process. The credible interval was not too different from the estimate obtained through simulation (three simulation runs produced 278-711, 278-640, and 278-640). (The upper bound estimates of 640 and 711 occur with ten and nine recaptures, respectively.)

The estimated posterior distribution for N.hat (Figure 4.2) can be obtained from a histogram plot of values from the jagsfit object. The abline() function adds a dotted vertical line for the true value.

```
hist(jagsfit$BUGSoutput$sims.list$N.hat, main="",
    xlab="Population estimates")
abline(v=N.true, lty=3, lwd=3)
```

The jagsfit object can be inspected in the Environment window. Note that sims.list arrays contain 3000 elements, which is the combined length of 1000 retained values (2000-n.burnin) from each of three chains. (When entering a

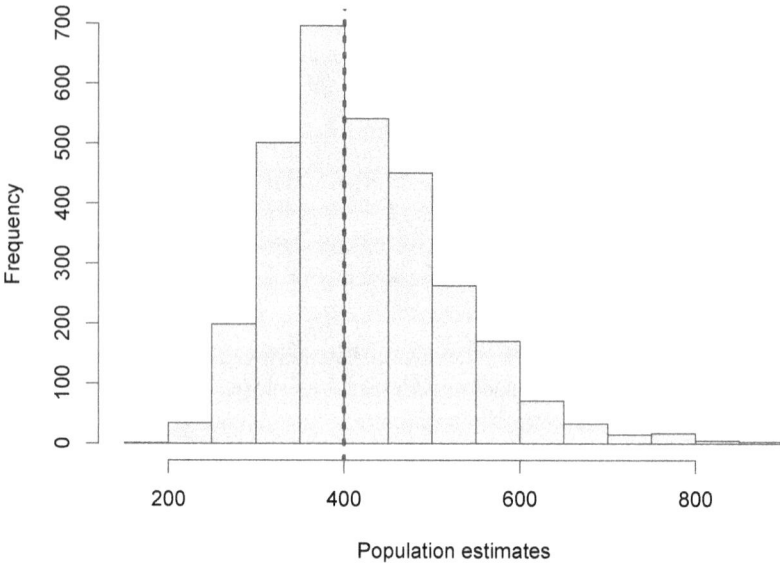

FIGURE 4.2 Estimated posterior distribution for population size from a two-sample mark-recapture study with a capture probability of 0.2; dotted vertical line indicates true value.

jagsfit object in the Source window or Console, typing $ after each level of the object will bring up an RStudio prompt for the next choice.)

The model results include pV, which is an estimate of the number of effective parameters in the model. In simple models, we can often count the number of nominal parameters, but the number of effective parameters may be less; for example, parameters may be correlated or constrained by an informative prior distribution (McCarthy, 2007). JAGS uses numerical methods to estimate pV. For this model, I would expect pV to be close to 1 for the estimate of N.hat; MarkedFraction is a calculated value and not a model parameter.

JAGS also provides an estimate of the Deviance Information Criterion (DIC), which is a relative measure of how well a model fits the data (McCarthy, 2007; Lunn et al., 2009). It is the Bayesian analog of Akaike's Information Criterion (AIC) which is widely used in a frequentist setting to compare alternative models. Like the AIC, the DIC results in a trade-off between model fit (likelihood) and complexity (number of parameters) (McCarthy, 2007). The DIC score is calculated using a measure of fit (deviance at the mean of the posterior distribution) and a measure of model complexity (pV); lower values are better when comparing alternative models. DIC scores should be interpreted with caution because of the difficulty in estimating pV for some models (Lunn et al., 2009).

This model may also be fitted by using MarkedFraction as the model parameter. It is easy to specify the uninformative prior (and initial values) in this case, as MarkedFraction is a proportion between 0 and 1. Now N.hat is a calculated value. One major advantage of Bayesian software is that we automatically get full information about parameters that are calculated functions of model parameters. In this case, the calculated value (N.hat) is of primary interest, rather than the actual model parameter MarkedFraction. Results for N.hat characterize its posterior distribution, which takes into account the prior information (none here) and the new data.

```r
rm(list=ls()) # Clear Environment

# Load necessary library packages
library(rjags)    # Package for fitting JAGS models from within R
library(R2jags)   # Package for fitting JAGS models. Requires rjags

# Arbitrary 'observed' values for analysis
N.true <- 400   # Population size
p.true <- 0.2 # Capture probability (fraction of population sampled)
n1 <- N.true * p.true # Caught and marked in first sample
n2 <- N.true * p.true # Caught and examined for marks in second sample
m2 <- n2 * p.true # Marked fish in second sample

# JAGS code
sink("TwoSampleCR.txt")
cat("
    model {

    # Priors
    MarkedFraction ~ dunif(0, 1)

    # Calculated value
    N.hat <- n1 /MarkedFraction

    # Likelihood
    # Binomial distribution for observed recaptures
    m2 ~ dbin(MarkedFraction, n2)
    }

    ",fill = TRUE)
sink()

# Bundle data
jags.data <- list("n1", "n2", "m2")
```

```
# Initial values.
jags.inits <- function(){ list(MarkedFraction=runif(1, min=0, max=1))}

model.file <- 'TwoSampleCR.txt'

# Parameters monitored
jags.params <- c("N.hat", "MarkedFraction")

# Call JAGS from R
jagsfit <- jags(data=jags.data, inits=jags.inits, jags.params,
                n.chains = 3, n.thin = 1, n.iter = 2000,
                n.burnin = 1000, model.file)
print(jagsfit, digits=3)
plot(jagsfit)
```

Try multiple runs using different values for p.true. In planning a mark-recapture study for a pond or small lake, how might you decide on a realistic goal for p.true?

4.3.2   Debugging code

Anyone who develops new code for fisheries data analysis will wrestle with the inevitable coding errors, not only in JAGS but also the R code that surrounds it. Let's begin by looking at a few examples of R coding errors:

```
rm(list=ls()) # Clear Environment

#1
x <- 2
lg(x) # Incorrect function name

#2
y <- 4
z <- x+y # Typo (should have been x*y)

#3
RandLength <- rnorm(1, 10, 2) # Correct but poorly documented
RandLength <- rnorm(10,2) # Sample size omitted
RandLength <- rnorm(1,10) # sd omitted so default used (incorrectly)
RandLength <- rnorm(n=1, mean=10, sd=2) # Correct and fully documented
```

```
#4
x2 <- rnorm(n=1000, mean=10, sd=5)
x3 <- rnorm(n=1000, mean=20, sd=10)
x4 <- x2+x3
mean(x4)
var(x4)
```

The first example is easy to resolve, as we get a helpful error message in the Console. Rechecking the code and doing an internet search for the correct function name are approaches for addressing this type of error. The second example, if our intent was to assign x*y to z, is much more dangerous. The code will run but will give the wrong answer because of a typing error. It is always essential to review new code, line by line. Having a colleague (fresh set of eyes) check code can also be productive. R code can often be executed line by line, and checking calculated values in the Environment window can help in finding this sort of logic error, at least for simpler analyses.

In the third example, we are simulating a random length observation, drawn from a normal distribution with a mean of 10 and standard deviation of 2. The first version of the code works correctly but is poorly documented. The second example runs but does not produce the intended result. We omitted the first argument and did not include parameter names, so R assumes incorrectly that the first argument is the sample size, and the second one is the mean (see *vector* RandLength in the Environment window). The third argument is missing, so R uses the default value (sd=1), which in this case is incorrect. Even if we wanted to use an sd of 1, it is better to specify it for clarity. RStudio is helpful in working with functions in that it displays the function arguments and any defaults, once the function name and "(" have been entered. The fourth version works as intended and makes clear the values that are being used.

The fourth example is coded correctly but illustrates another way of checking code. Simulation makes it easy to use a large sample size, to check sample statistics for x4. We verify that the mean is close to the expected value (10+20). The variance of a sum (of independent variables) equals the sum of the variances, so we confirm that the sample variance is close to the expected value (25+100).

How many errors do you find in the following section of code? Review line by line and correct errors before executing the code. Once the code is working correctly, confirm that the sample median is very close to the expected population size:

```
rm(list=ls()) # Clear Environment

# Two-sample population estimate
N.true <- 400   # Arbitrary assumed population size
p.true <-- 0.2 # Capture probability (fraction of population sampled)
n1 <- N.true X p.true # Caught and marked in first sample
n2 <- N.true * p.true # Caught and examined for marks in second sample

Reps <- 1000
m2.vec <- rbinom(n=Rep, prob=n1/N.true, size=n2)
N.hat <- n1 * n2 / m2.vec
hist(N.hat, xlab="Population estimate", main="")
mean(N.hat)
quantile(N.hat, probs=c(0.5), na.rm=TRUE))   # Sample median
```

Debugging JAGS code can be more challenging. There is no option for executing
single lines of code as is possible with R. It is usually a matter of fixing one
error at a time until the code runs, then checking and testing the code to be
confident that it is working correctly. Consider the following (correctly coded)
example for a tagging study to estimate the exploitation rate (fraction of the
population harvested). Our simulation generates the single observation for the
number of tags returned (e.g., over a year). This observation is drawn from a
binomial distribution, and we use the same JAGS function (dbin()) as in the
population estimate examples:

```
rm(list=ls()) # Clear Environment

# Arbitrary 'observed' values for analysis
N.tagged <- 100
Exp.rate <- 0.4
Tags.returned <- rbinom(n=1, prob=Exp.rate, size=N.tagged)

# Load necessary library packages
library(rjags)   # Package for fitting JAGS models from within R
library(R2jags)  # Package for fitting JAGS models. Requires rjags

# JAGS code
sink("ExpRate.txt")
cat("
    model {

      # Priors
```

```
    Exp.rate.est ~ dunif(0,1)

    # Likelihood
    Tags.returned ~ dbin(Exp.rate.est, N.tagged)
}
    ",fill = TRUE)
sink()

# Bundle data
jags.data <- list("N.tagged", "Tags.returned")

# Initial values.
jags.inits <- function(){ list(Exp.rate.est=runif(n=1, min=0, max=1))}

model.file <- 'ExpRate.txt'

# Parameters monitored
jags.params <- c("Exp.rate.est")

# Call JAGS from R
jagsfit <- jags(data=jags.data, inits=jags.inits, jags.params,
                n.chains = 3, n.thin = 1, n.iter = 2000,
                n.burnin = 1000, model.file)
print(jagsfit, digits=3)
plot(jagsfit)
```

Begin by running the code as is, then try substituting each of the following (incorrect) lines, one at a time.

- `ExpRate.est ~ dunif(0,1)`
- `Exp.rate.est ~ unif(0,1)`
- `Tags.returned ~ dbin(Exp.rate.est)`

For example, the incorrect line of code could be:

```
ExpRate.est ~ dunif(0,1)   #Exp.rate.est ~ dunif(0,1)
```

Rerunning the code (click on RStudio icon to Rerun the previous code region) produces an error message pointing to line 8 (counting from cat("). JAGS correctly reports that the variable Exp.rate.est has not been previously defined. This is correct, but the error is actually on line 5, where the intended parameter name (Exp.rate.est) was mistyped. The second error is easy to diagnose, because it specifies that the "unif" distribution on line 5 is unknown. The third error indicates that a discrete-valued parameter for function dbin() is missing. This makes sense because the size of the experiment (N.tagged) was omitted.

As with R code, the first step is always to review the code line by line. Fitting a model to simulated data is always useful (even if there are field observations) because the expected results are known. Another strategy is to start with the simplest possible analysis, then add complexity one piece at a time. For example, Kéry and Schaub (2012) discuss Cormack-Jolly-Seber (Royle, 2008) models for estimating survival rate, beginning with constant parameters then gradually adding more complexity (e.g., models that allow for time and individual variation).

Skill in debugging comes with experience. The above strategies should be useful in resolving coding errors.

4.3.3   Judging convergence

The MCMC process is iterative, so it is important to ensure that the number of updates (iterations) is sufficient to achieve convergence. One practical approach is to fit the model with some arbitrary (but reasonably large) number of updates, then refit the model using larger numbers of updates. Results that are stable suggest that convergence has been achieved. The convergence statistic Rhat (Gelman and Rubin, 1992; Brooks and Gelman, 1998; Lunn et al., 2012) is also generally reliable as a measure of convergence. Especially for models for which convergence appears to be slow, it can also be helpful to view "trace" plots that show how estimates change as updating occurs (Kéry, 2010). Add the following lines of code to the two-sample mark-recapture model used in Section 4.3:

```
par(mfrow=c(3,1)) # Multi-frame plot, 3 rows, 1 col
matplot(jagsfit$BUGSoutput$sims.array[,,1], type = "l",
        ylab="Marked fraction")
matplot(jagsfit$BUGSoutput$sims.array[,,2], type = "l", ylab="N hat")
matplot(jagsfit$BUGSoutput$sims.array[,,3], type = "l", ylab="Deviance")

jagsfit.mcmc <- as.mcmc(jagsfit) # Creates an MCMC object for plotting
plot(jagsfit.mcmc) # Trace and density plots
```

The first approach is based on an example from Kéry (2010). We use the par() graphical function to set up a multi-frame plot (by row), with three rows and one column. This sets up a single RStudio plot window to contain three plots. The next three lines use the built-in matplot() function to plot columns of a matrix as overlaid lines. The Environment window can be viewed to see the structure of the sims.array, a matrix of type num[1:1000, 1:3, 1:3]. The first dimension is for the 1000 retained updates, the second is for chains 1-3, and the third is for the variable (MarkedFraction, N.hat, Deviance). Omitting the ranges for updates and chains results in a plot containing all retained updates, with a different line color for each chain. We could also specify a subset of the range to see the initial pattern (e.g., … sims.array[1:100,,1], …). For this simple model, the estimates converge almost immediately, so there is no obvious transitory pattern. Note that there is a lot of wasted space and it may be necessary to increase the size of the Plots window to see the plots adequately.

The final two lines provide a more automated way of plotting the sequence of values (trace) and density (Figure 4.3), using methods contained in the **coda** (Plummer et al., 2024) package (loaded automatically by rjags). The as.mcmc() function converts the original output object (jagsfit) into an MCMC object. The plot function here produces a nicely formatted set of trace and density plots that are very helpful in confirming convergence. If convergence has been achieved, the trace plots should appear to be a "grassy lawn" or "fuzzy caterpillar" without any trend, with overlapping (fully mixed) chains. Density plots show the estimated posterior distributions, and given convergence, should have an essentially identical pattern for the three chains. Convergence occurs very quickly here, but we will see in later chapters with more complex models that convergence can sometimes be more difficult to achieve (see, especially, Section 8.2).

4.3.4 What is a likelihood?

As mentioned above, Bayesian software uses the likelihood and prior information to estimate the posterior distribution. But what exactly is a likelihood? It is an expression or function that is used to estimate the *likelihood* of obtaining the sample observation(s), given some specific value for each model parameter. As an example, let's return to the second version of the mark-recapture analysis, with the single model parameter MarkedFracton (the fraction of the population that is marked). We use the binomial distribution to model the number of recaptures. The likelihood function for the binomial is the same as the probability distribution: $L(p|n, k) = \binom{n}{k} p^k (1 - p)^{n-k}$, except that here, we estimate the likelihood of a certain value for p, given k successes in a trial of size n. We can look at likelihoods using the same parameter values as in the simulation:

```
rm(list=ls()) # Clear Environment

# Arbitrary 'observed' values for analysis
```

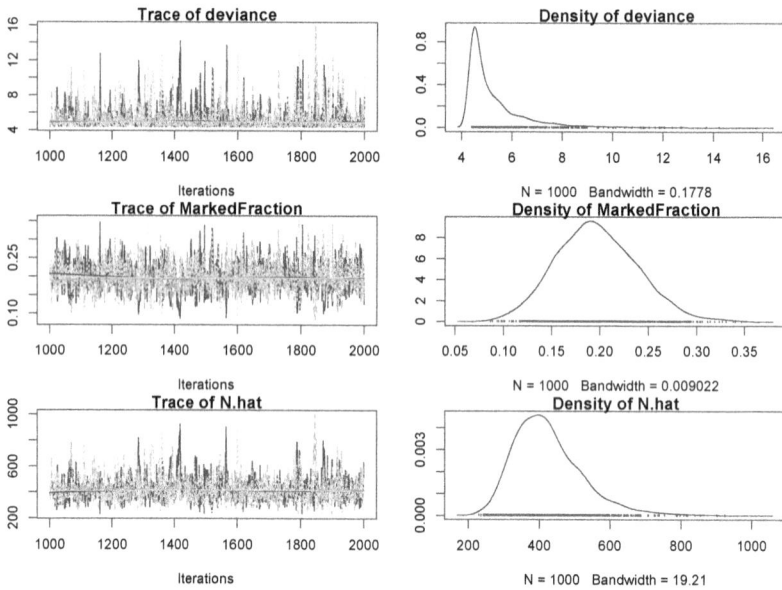

FIGURE 4.3 Trace and density plots for the two-sample mark-recapture model, using methods contained in the coda package.

```
N.true <- 400   # Population size
p.true <- 0.2 # Capture probability (fraction of population sampled)
n1 <- N.true * p.true # Caught and marked in first sample
n2 <- N.true * p.true # Caught and examined for marks in second sample
m2 <- n2 * p.true # Marked fish in second sample

# Calculate likelihood for specific value of p
p <- 0.2
choose(n2,m2) * p^m2 * (1-p)^(n2-m2) # Likelihood for single value of p

dbinom(x=m2, size=n2, prob=p)
# x successes, built-in function to obtain same result

# Create vector of possible p values, determine likelihood for each
p.vec <- seq(from=0.02, to=1, by=0.02)
l.vec <- choose(n2,m2) * p.vec^m2 * (1-p.vec)^(n2-m2)
par(mfrow=c(1,1)) # Reset plot frame
plot(p.vec, l.vec, ylab="Likelihood", col="red")

# Examine change in likelihood at higher number of recaptures
new.m2 <- 18
```

```
new.l.vec <- choose(n2,new.m2) * p.vec^new.m2 * (1-p.vec)^(n2-new.m2)
points(p.vec, new.l.vec, col="blue")
```

The likelihood calculation at p=0.2 uses the choose() function, to determine how many ways m2=16 recaptures can be drawn from a sample of size 80. Type choose(n2,m2) in the Console to see that it is a very large number! The remainder of the expression determines the probability of 16 successes and 80-16 failures. We can obtain the same result from the built-in function dbinom(), although it seems more instructive to write out the full expression.

The calculated likelihood (0.11) is really only of interest as a relative value, compared to other potential values for p. We use the seq() function to create a vector of potential p values, and obtain a vector of likelihoods from the p vector. Plotting the points shows that the likelihood peaks sharply at 0.2 (not surprisingly, since that is the value that generated the 16 recaptures) and decreases to near 0 at about 0.08 and 0.32.

We have assumed thus far that we obtained the expected number of recaptures (80*p), but in reality, the number of recaptures would be a binomially-distributed random variate. A different observed value for m_2 would shift the likelihood function. The next lines of code calculate the likelihood for a slightly higher number of recaptures. The points() function adds the new likelihood values to the plot, using the col="blue" plot option. The new values are shifted to the right because the number of recaptured fish was higher.

If multiple independent samples are taken, the combined likelihood is a product. For our two assumed samples of 80 fish, with 16 recaptures in one and 18 in the other, the combined likelihood at each level of p would be the product l.vec * new.l.vec.

We wrote out the expression for the likelihood for this example, but that is not needed when using the BUGS language. We need only specify the function name for the distribution describing our sample data. JAGS does the behind-the-scenes work to provide the code for the likelihood function. Working at a higher level allows us to focus on the study design and assumptions, which guide us to our choice of sample distribution.

4.3.5   Deviance as a measure of model fit

Likelihood calculations were introduced in Section 4.3.4. Here we show two related measures of model fit (log-likelihood and deviance) using the same mark-recapture example. There can be computational advantages to working with the log-likelihood compared to the likelihood (McCarthy, 2007).

```
rm(list=ls()) # Clear Environment

# Arbitrary 'observed' values for analysis
N.true <- 400   # Population size
p.true <- 0.2 # Capture probability (fraction of population sampled)
n1 <- N.true * p.true # Caught and marked in first sample
n2 <- N.true * p.true # Caught and examined for marks in second sample
m2 <- n2 * p.true # Marked fish in second sample

# Create vector of possible p values, determine likelihood for each
p.vec <- seq(from=0.02, to=1, by=0.02)
l.vec <- choose(n2,m2) * p.vec^m2 * (1-p.vec)^(n2-m2)
plot(p.vec, l.vec, ylab="Likelihood", col="red")

log_l.vec <- log(l.vec) # Log-likelihood
plot(p.vec, log_l.vec, ylab="log-Likelihood", col="red")
dev.vec <- -2 * log_l.vec # Deviance (-2 * ln_L)
plot(p.vec, dev.vec, ylab="Deviance", col="red")
```

Compared to the likelihood, the log-likelihood curve has a different shape but the same peak at p=0.2 (the most likely value given 16 successes in a trial size of 80). The deviance is obtained by multiplying the log-likelihood by -2. The plotted pattern is now inverted so that now the curve has a *minimum* at p=0.2, i.e., a larger value for deviance indicates a poorer fit (McCarthy, 2007). JAGS uses the deviance to obtain parameter estimates, but it is nice to know that we would obtain the same estimates from either the maximum likelihood or log-likelihood or the minimum deviance (McCarthy, 2007).

4.3.6   Model checking

It is easy to fit the correct model to simulated data, but that is not the case for real (field) data. Any model fit to field data should be viewed as an approximation of the underlying ecological processes. We cannot prove that the model is correct, but we can examine whether it is a useful approximation. One method for doing this is the posterior predictive check (Gelman and Hill, 2007; Kéry, 2010; Kéry and Schaub, 2012; Conn et al., 2018). This method compares a measure of fit for replicated data generated using the fitted model compared to the observed data. If the measure of fit (e.g., sum of squares or chi-square) is markedly different (usually higher) for the observed data, that suggests that the observed data contain some features not captured by the fitted model.

Our first example uses simulated catch data from a Poisson distribution, as might be obtained by standardized electrofishing or trawling. Keep in mind that these are our "observed" catches, to be compared with replicate data obtained by simulation. I arbitrarily set the mean for the Poisson distribution at 7 and the sample size at 30. Before viewing the extended JAGS code that includes the posterior predictive check, let's examine a version that simply fits the model:

```r
# Fit model to catch data from Poisson distribution

rm(list=ls()) # Clear Environment

N <- 30
mu <- 7
Catch <- rpois(n=N, lambda=mu) # Drawn from Poisson distribution
Freq <- table(Catch)  # Distribution of simulated catches
barplot(Freq, main="", xlab="Catch", ylab="Frequency")

# Load necessary library packages
library(rjags)
library(R2jags)

sink("PoissonFit.txt")
cat("
model {

# Prior
  lambda.est ~ dunif(0, 100)

# Likelihood
    for(i in 1:N) {
      Catch[i] ~ dpois(lambda.est)
      } #y
}
    ",fill=TRUE)
sink()

# Bundle data
jags.data <- list("N", "Catch")

# Initial values
jags.inits <- function(){ list(lambda.est=runif(n=1, min=0, max=100))}

model.file <- 'PoissonFit.txt'
```

```
# Parameters monitored
jags.params <- c("lambda.est")

# Call JAGS from R
jagsfit <- jags(data=jags.data, jags.params, inits=jags.inits,
                n.chains = 3, n.thin=1, n.iter = 10000,
                model.file)
print(jagsfit)
plot(jagsfit)
```

There are only a few lines of JAGS code. We use an uninformative uniform prior distribution for the single Poisson parameter lambda.est. The catches are assumed (correctly) to be drawn from a Poisson distribution, and the likelihood calculations use the Poisson distribution function dpois(). The initial value is obtained from the same uniform distribution as the prior. JAGS returns the estimated mean, which varies around the true value due to the modest sample size. Try rerunning the code using a large sample size to see the improvement in the sample catch plot and the estimated mean.

Next we add a few lines of code to carry out the posterior predictive check:

```
# Fit model to catch data from Poisson distribution
# Code includes posterior predictive check

rm(list=ls()) # Clear Environment

N <- 30
mu <- 7
Catch <- rpois(n=N, lambda=mu) # Drawn from Poisson distribution
Freq <- table(Catch)   # Distribution of simulated catches
barplot(Freq, main="", xlab="Catch", ylab="Frequency")

# Load necessary library packages
library(rjags)
library(R2jags)

sink("PPC_Example.txt")
cat("
model {

# Prior
 lambda.est ~ dunif(0, 100)
```

```
# Likelihood
    for(i in 1:N) {
      Catch[i] ~ dpois(lambda.est)
      Catch.new[i] ~ dpois(lambda.est) # Replicate new data for PPC
      # Chi-square discrepancy values for observed and replicate data
      chi.obs[i] <- pow(Catch[i]-lambda.est,2)/lambda.est
      chi.new[i] <- pow(Catch.new[i]-lambda.est,2)/lambda.est
      } #y
  Total.obs <- sum(chi.obs[])
  Total.new <- sum(chi.new[])
  Bayesian.p <- step(Total.new - Total.obs)
}
    ",fill=TRUE)
sink()

# Bundle data
jags.data <- list("N", "Catch")

# Initial values
jags.inits <- function(){ list(lambda.est=runif(n=1, min=0, max=100))}

model.file <- 'PPC_Example.txt'

# Parameters monitored
jags.params <- c("lambda.est", "Bayesian.p")

# Call JAGS from R
jagsfit <- jags(data=jags.data, jags.params, inits=jags.inits,
                n.chains = 3, n.thin=1, n.iter = 10000,
                model.file)
print(jagsfit)
plot(jagsfit)
```

Within the likelihood looping section, we draw new replicate observations from the posterior distribution (Catch.new), and calculate chi-square values for observed and replicate data. The Bayesian p-value is obtained from the step() function, which returns a 0 if its argument (difference in total chi-square values) is negative and a 1 if 0 or greater. Thus the variable Bayesian.p has a sequence of 0 and 1 values, and the mean provides the proportion of updates for which Total.new was the larger of the two values. The Bayesian p-values tend to be quite variable; ten trials produced values of 0.34, 0.42, 0.18, 0.16, 0.32, 0.30, 0.32, 0.67, 0.69, and 0.18. We would expect the total chi-square for observed and replicate data to be the same on average, as the fitted model is known to be correct (i.e., our "observed" catches and the replicate data sets are drawn from Poisson distributions). Keep in mind that, when analyzing a real data set, there will be a single Bayesian p-value rather than the replicated values shown above. Simulations mimicking the field study are very helpful in interpreting that single value.

Next, consider a case where the "observed" data are obtained from something other than a Poisson distribution. The negative binomial distribution (Section 3.1.5) is a good choice because it can approximate a Poisson or have a more skewed distribution, depending on the value for the overdispersion parameter k. Here we use the same mean as in the Poisson example and set k=1 (variance much larger than mean). The JAGS code for model fitting remains unchanged but we replace the Poisson simulation code with the following:

```
# Fit model to negative binomial data
# Code includes posterior predictive check

rm(list=ls()) # Clear Environment

N <- 30
mu <- 7
k=1
variance <- mu+(mu^2)/k
Catch <- rnbinom(n=N, mu=mu, size=k) # Catches drawn from neg binomial
Freq <- table(Catch)  # Distribution of simulated counts
barplot(Freq, main="", xlab="Count", ylab="Frequency")
```

Now Bayesian p-values are consistently extreme (0.00 for five replicate trials). The total chi-square for the replicate data sets (which *are* Poisson distributed) is always less than for the observed data (which are not). The fitted Poisson model lacks the flexibility to mimic the skew of the observed catch distribution. More intermediate Bayesian p-values are only obtained if k takes on larger values, such that the negative binomial approximates a Poisson distribution (Figure 4.4).

Kéry and Schaub (2012) list some concerns about the posterior predictive check. One is that the observed data are used twice: in fitting the model and in producing the Bayesian p-value. Another is that the method is qualitative, in that there is not an objective p-value range that indicates an acceptable fit. Despite those concerns, lack of fit can be established by extreme p-values. Kéry (2010) suggested poor fit was indicated by values close to 0 or 1; Conn et al. (2018) suggested p-values less than 0.05 or greater than 0.95. These recommmendations appear to be consistent with the simulation results for k<10 in Figure 4.4. To better understand this pattern in Bayesian p-values, try

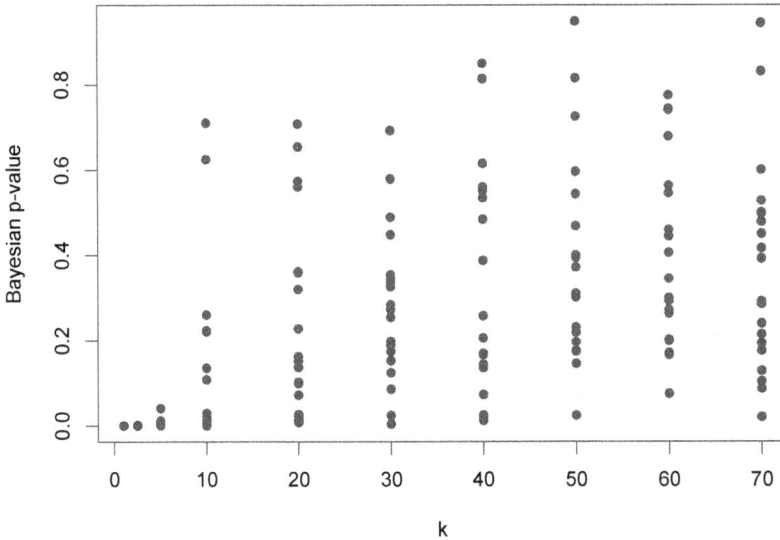

FIGURE 4.4 Bayesian p-value for fitting a Poisson model to simulated data from a negative binomial distribution with different values for overdispersion parameter k but fixed mean=7. Twenty replicate simulations were done for each value of k.

the following code as a simple graphical way to compare a Poisson distribution with negative binomial distributions generated using a range of values for k. The range() function provides a common x-axis for the two plots to facilitate comparisons.

```
x.p <- rpois(n=1000, lambda=7)
k <- 1 # Overdispersion parameter
x.np <- rnbinom(n=1000, size=k, mu=7)
x.bound <- range(x.p, x.np)
par(mfrow=c(1,2))
hist(x.p, main="", xlim=x.bound, xlab="Poisson")
hist(x.np, main="", xlim=x.bound, xlab="Negative binomial")
```

My experience has been that posterior predictive checks can be a useful way of detecting lack of fit. Simulation is helpful in gaining experience with the approach, because it is straightforward to obtain Bayesian p-values for correct and incorrect models. Also, simulation of a specific study design and model can be used to determine the range of p-values that would be considered extreme (Conn et al., 2018).

4.3.7   Model selection

It is often the case that more than one model may be considered as plausible. For example, survey catches could be modeled using either a Poisson or negative binomial distribution. A model for a tag-return study could include separate parameters for tag-reporting rate from commercial and recreational sectors, or just a single parameter if the reporting rates were similar. A third example is a model for survey catch rate, which might include covariates such as location, year, depth, or water temperature. Determining which covariates affected catch rate would make it possible to produce an improved survey index that was adjusted for environmental variation. For example, were low survey catches in the current year due to anomalously low water temperatures on sampling dates or to a real decline in abundance? The suite of candidate models should be chosen before data analysis begins (Burnham and Anderson, 1998) to avoid the tendency to examine the study results and to choose candidate models that match observed patterns. For example, observing low recruitment in a recent year might spur us to search for environmental covariates with a similar temporal pattern.

Choosing a preferred model among the suite of candidates (model selection) is a vast and complex topic (Link and Barker, 2010), and there is no consensus among statisticians as to the best approach (Kéry and Schaub, 2012). Kéry and Schaub (2012) suggest that one practical approach is to decide on a model that is biologically plausible and stick to that. They also sometimes eliminate candidate parameters that have credible intervals that include zero, similar to a backward stepwise regression. Hooten and Hobbs (2015) recommend model selection based on predictive ability, either in-sample (data used in fitting the model) or out-of-sample (new data). Out-of-sample validation is considered the gold standard (Hooten and Hobbs, 2015). The main disadvantage of that approach is the requirement for additional data not used in fitting the model. Doll and Jacquemin (2019) demonstrate and provide Stan (Gelman et al., 2015) code for additional methods of model selection including WAIC (Watanabe-Akaike or widely applicable information criterion: Watanabe (2010)). WAIC is not calculated automatically by JAGS, but, unlike DIC, it can be used in hierarchical and mixture models (Gelman et al., 2014; Hooten and Hobbs, 2015).

We illustrate here a simpler approach of separately fitting candidate models and comparing DIC scores. Our "observed" data are simulated counts from a negative binomial distribution. Next, we fit Poisson and negative binomial candidate models to the same observed data and compare DIC scores. The negative binomial distribution in JAGS uses the probability of success (p) and size (our overdispersion parameter, k). We return the estimated mean for the negative binomial distribution, which is calculated internally using p.est and k.est.

```
# DIC comparison for model selection, using "observed" data
# from a negative binomial distribution

rm(list=ls()) # Clear Environment

N <- 30
mu <- 7
k=1
variance <- mu+(mu^2)/k
Count <- rnbinom(n=N, mu=mu, size=k) # Drawn from negative binomial
Freq <- table(Count)  # Distribution of simulated counts
barplot(Freq, main="", xlab="Count", ylab="Frequency")

# Load necessary library packages
library(rjags)
library(R2jags)

sink("PoissonFit.txt")
cat("
model {

# Prior
 lambda.est ~ dunif(0, 100)

# Likelihood
    for(i in 1:N) {
       Count[i] ~ dpois(lambda.est)
       } #y
}
    ",fill=TRUE)
sink()

# Bundle data
jags.data <- list("N", "Count")

# Initial values
jags.inits <- function(){ list(lambda.est=runif(n=1, min=0, max=100))}

model.file <- 'PoissonFit.txt'

# Parameters monitored
jags.params <- c("lambda.est")

# Call JAGS from R
```

```
jagsfit <- jags(data=jags.data, jags.params, inits=jags.inits,
                n.chains = 3, n.thin=1, n.iter = 10000,
                model.file)
print(jagsfit)
#plot(jagsfit)


Poisson.DIC <- jagsfit$BUGSoutput$DIC

sink("NBFit.txt")
cat("
model {

# Prior
 p.est ~ dunif(0, 1)
   # JAGS uses p (probability of success) for negative binomial
 k.est ~ dunif(0, 1000)
 mu.est <- k.est*(1-p.est)/p.est

# Likelihood
    for(i in 1:N) {
      Count[i] ~ dnegbin(p.est, k.est)
      } #y
}
    ",fill=TRUE)
sink()

# Bundle data
jags.data <- list("N", "Count")

# Initial values
jags.inits <- function(){ list(p.est=runif(n=1, min=1E-6, max=1),
                               k.est=runif(n=1, min=0, max=1000))}

model.file <- 'NBFit.txt'

# Parameters monitored
jags.params <- c("mu.est", "k.est")

# Call JAGS from R
jagsfit <- jags(data=jags.data, jags.params, inits=jags.inits,
                n.chains = 3, n.thin=1, n.iter = 10000,
                model.file)
print(jagsfit)
#plot(jagsfit)
```

```
NB.DIC <- jagsfit$BUGSoutput$DIC
Poisson.DIC - NB.DIC
```

The difference in DIC scores (printed to the Console for convenience) ranged from 50.4 to 90.4 in five trials using the default simulation settings. The substantially higher DIC score for the (incorrect) Poisson model is clear evidence of a poorer fit. Burnham and Anderson (1998) suggested that an alternative model with an AIC score within 1-2 of the best model merits consideration, whereas a difference of 3-7 suggests much less support. Spiegelhalter et al. (2002) noted that the Burnham and Anderson (1998) criteria appear to work well for DIC scores.

Comparing DIC scores worked well for this simple (non-hierarchical, non-mixture) case. Also, using simulated counts from a negative binomial distribution is convenient because we can adjust the overdispersion parameter (k) to obtain a more "Poisson-like" distribution. At what value for k do you obtain similar DIC scores for the two models? Observe the change in shape of the count frequency distribution as you vary k.

4.4 Exercises

1. Estimate uncertainty around a two-sample mark-recapture estimate N.hat using the simulation and Bayesian approaches for a capture probability of 0.4. How similar are the results, and how do they compare to results using the original capture probability of 0.2?

2. Robson and Regier (1964) provided practical advice on acceptable levels of accuracy and precision. They suggested that estimates within 10% of the true value be acceptable for research studies, versus 25% for management and 50% for preliminary studies. Bayesian credible intervals could be used as a proxy for their accuracy and precision targets. What capture probabilities (approximate) would correspond to those targets for the mark-recapture example?

3. How many errors can you locate before running the following code? Fix all errors that you spotted, then run the code to fix any remaining errors.

```r
rm(list=ls()) # Clear Environment

# Load necessary library packages
library(rjags)    # Package for fitting JAGS models from within R
library(R2jags)   # Package for fitting JAGS models. Requires rjags

# Arbitrary 'observed' values for analysis
Reps <- 20
AveC <- 3
Count <- rpois(n=Reps lambda=AveC)

# JAGS code
sink("PoissonSim.txt")
cat("
    model {

    # Priors
    AveC.est ~ dunif(0, MaxCount)

    # Likelihood
    for (i in 1:Rep){
    Count[i] ~ dpois(AveCest)
    } #i
}
    ",fill = TRUE)
sink()

# Bundle data
jags.data <- list("Count")

# Initial values.
jags.inits <- function(){ list(AveC.est=runif(n=1, min=0, max=100))}

model.file <- 'PoisonSim.txt'

# Parameters monitored
jags.params <- c("AveC.est")

# Call JAGS from R
jagsfit <- jags(data=jags.data, inits=jags.inits, jags.params,
                n.chains = 3, n.thin = 1, n.iter = 2000,
                n.burnin = 1000, model.file)
print(jagsfit, digits=3)
plot(jagsfit)
```

4. Consider a mark-recapture study with $n_1 = 60$ and two independent samples for recaptures ($n_2{=}90$, $m_2{=}40$; $n_3{=}55$, $m_3{=}30$). Use a vector for p at 0.01 intervals, and produce a vector and plot of joint likelihoods (vector given the name "both"). Assume that only the initial sample (n_1) is marked; samples n_2 and n_3 provide two independent snapshots of the marked fraction. Explain in detail how p[which.max(both)] works, and what it produces.

5. Use the approach from Section 4.3.7 to compare model fits for a more "Poisson-like" negative binomial distribution (e.g., k=10).

5

Abundance

Information about abundance can be invaluable for making policy decisions. For fish populations that support commercial or recreational fishing, that information could determine the appropriate level of harvest. For rare or declining populations, the information might be used to develop a restoration plan. It is rarely possible to conduct a census (counting every individual in a target population), so models are used to account for incomplete sampling. For example, we would expect to capture only a fraction of the target population even if we sampled the entire shoreline of a lake using electrofishing. The probability of capturing an individual fish during a sampling event is referred to as the capture probability. It can also be thought of as the sampling fraction, or the fraction of the target population that is caught or counted (Williams et al., 2002). In some cases, the probability represents detection rather than capture, for example, the probability of detecting an individual fish with a transmitter during a search. Estimating capture or detection probabililties allows us to correct for imperfect sampling in order to estimate population parameters of interest (Williams et al., 2002).

In this chapter, we consider a few simple field approaches for estimating population size. These methods can be used in freshwater or marine systems, but they do require that the study area be closed (no migration in or out, no recruitment (additions through reproduction), and no mortality for the duration of the study). Later chapters will consider open models that jointly estimate population abundance and mortality over longer time frames.

5.1 Removal

The removal method is a good approach for estimating population size within a study area when a substantial fraction of the study population can be captured in a single sample. One common example is use of backpack electrofishing in streams that can be blocked on both ends to provide a temporarily closed system. Fish captured in each sample (or "pass") are held out temporarily and the declining catches from the diminishing population provide the data for estimating abundance. Consider a case with a true (study area) population of 100 and a capture probability of 0.4. The expected sequence of catches would be 40 (100 * 0.4), 24 ((100-40) * 0.4), 14.4 ((60-24)*0.4), etc. A model fit to these data provides estimates of initial population size and capture probability that best mimic the pattern of catches. The fit would be very good if we used the above expected values, but it is more challenging (and realistic) when the catch sequence includes random variation.

We illustrate the removal method using a simulation to generate sample data. Here the true values are arbitrary, but roughly similar to estimates obtained by Bryant (2000) for Alaska streams. If planning a field study, your assumed population level should be based on prior work or literature review. The assumed capture probability should also be realistic, but it is a bit different from population size in that it is under the investigator's control. Thus, a better estimate can be achieved by increasing sampling effort; for example, by using more electrofishing units or setting more traps. The other study design variable controlled by the investigator is the number of samples. Here we use five removal samples, which should provide high quality results.

```
rm(list=ls()) # Clear Environment

# Generate simulated removals
N.removals <- 5 # Number of removal samples
N.true <- 100    # True population size
CapProb <- 0.4   # Capture probability
N.remaining <- N.true # Initialize N.remaining, before removals begin
Catch <- numeric() # Creates empty vector for storing catch data
for (j in 1:N.removals){
  Catch[j] <- rbinom(1, N.remaining, CapProb)
  N.remaining <- N.remaining - Catch[j]
  } #j
```

Running the above lines of code generates the sequence of simulated catches. (Note that these lines of simulation code would be replaced by an assignment statement (e.g., Catch <- c(14, 4, 6, 2, 3)) if we were analyzing data from an actual field study). A 'for' loop is used to step through the simulated sampling sequence. For each value of j, we generate a binomially-distributed random catch using the rbinom() function, then remove that number of fish from the remaining population. One example run produced a catch vector of 36, 25, 20, 6 and 6. The high assumed value for capture probability produces very good data; thus, these values are close to what would be expected. Try running the code using other values for capture probability. Is there a lower bound below which you would not expect the method to work well?

The next step is to fit the removal model using JAGS. The prior distribution for capture probability is straightforward (uniform 0-1), but the choice is a bit more subjective for population size (N.est). We know that N.est cannot be less than the total catch, so we can use a slightly informative prior (with an arbitrary high value for the upper bound). Results from test runs were similar when using 0 as the lower bound, and when different values were used for the upper bound. Another alternative that provides similar results is the vague lognormal prior used in Section 4.3.1.

```
# Load necessary library packages
library(rjags)    # Package for fitting JAGS models from within R
library(R2jags)   # Package for fitting JAGS models. Requires rjags

# JAGS code for fitting model
sink("RemovalModel.txt")
  cat("
model{

# Priors
 CapProb.est ~ dunif(0,1)
 N.est ~ dunif(TotalCatch, 2000)

 N.remaining[1] <- trunc(N.est)
 for(j in 1:N.removals){
     Catch[j]~dbin(CapProb.est, N.remaining[j]) # jth removal
     N.remaining[j+1] <- N.remaining[j]-Catch[j]
   } #j

}
    ",fill=TRUE)
  sink()

    # Bundle data
  TotalCatch <- sum(Catch[])
  jags.data <- list("Catch", "N.removals", "TotalCatch")


# Initial values
  jags.inits <- function(){ list(CapProb.est=runif(1, min=0, max=1),
                             N.est=runif(n=1, min=TotalCatch,
                             max=2000))}

  model.file <- 'RemovalModel.txt'
```

```
# Parameters monitored
jags.params <- c("N.est", "CapProb.est")

  # Call JAGS from R
jagsfit <- jags(data=jags.data, jags.params, inits=jags.inits,
                n.chains = 3, n.thin = 1, n.iter = 40000,
                model.file)
print(jagsfit)
plot(jagsfit)
```

JAGS requires an integer size variable in dbin(), so we use the trunc() function to truncate the population estimate at time 1 (N.remaining[1]). The looping code is the same as in the simulation, except that JAGS does not allow a value to be repeatedly changed in a loop, so we set up the remaining population size as a vector. We provide to JAGS the simulated catch vector, the number of removal samples, and the total catch. Initial values are generated using the same distributions as the priors. We monitor the estimates of population size and capture probability.

Initial runs using fewer updates occasionally resulted in poor convergence, so n.iter was increased to 40000. Removing n.burnin uses the default of discarding half the updates. Running the simulation and analysis code multiple times for these simulation parameters shows that the estimates are consistently good, with narrow credible intervals. An example using fixed catch values is included as Appendix A.1.

As a fishery biologist, you could use this combination of simulation and analysis to plan your field study. You could look at different choices for the two design variables (number of removal samples, capture probability) to decide what would produce reliable results. White et al. (1982) suggest that capture probability should be at least 0.2, with better results obtained at 0.4 or higher with 3-6 removal samples. Try runs using the above code, modified to test different numbers of removal samples and assumed capture probabilities. It is important to acknowledge that this is an optimistic case, in that the data exactly follow binomial sampling. In a real field study, the model will be an approximation so any design choices should be viewed as lower bounds. For example, if four removal samples appear to be sufficient, consider that a minimum or perhaps low. Another consideration is that we have made the simplifying assumption that capture probability is constant across samples. Bryant (2000) found that capture probability was generally constant when using traps, but it may decline with disruptive methods such as electrofishing or seining, or if vulnerability varies among individuals (Mäntyniemi et al., 2005). If capture probability declines as additional samples are taken, a more complex function would be needed for capture probability.

5.2 Mark-recapture

Population size within a study area can also be estimated through a two-sample mark-recapture experiment. Fish captured in the first sample are given a tag or mark, and the population estimate is based on the fraction of marked fish in the second sample. We change one line of the code from the example in Section 4.3.1 to produce a random number of recaptured fish rather than the expected value. Now the combined code (simulation plus analysis) can be rerun multiple times to build intuition about the reliability of the experimental results. This is especially useful when the assumed population size or capture probability is low. The true value can be compared to the 95% credible interval (2.5 and 97.5 percentiles, printed to the Console) or 80% credible interval shown in the plots. You can look at either the marked fraction or N.hat, as they covary (low estimate for marked fraction produces high estimate for N.hat).

```
rm(list=ls()) # Clear Environment

# Simulate experiment
N.true <- 400  # Population size
p.true <- 0.2 # Capture probability (fraction of population sampled)
n1 <- N.true * p.true # Caught and marked in first sample
n2 <- N.true * p.true # Caught and examined for marks in second sample
m2 <- rbinom(n=1, size=n2, prob=p.true) # Marked fish in second sample

# Load necessary library packages
library(rjags)   # Package for fitting JAGS models from within R
library(R2jags)  # Package for fitting JAGS models. Requires rjags

# JAGS code
sink("TwoSampleCR.txt")
cat("
    model {

    # Priors
    MarkedFraction ~ dunif(0, 1)

    # Calculated value
    N.hat <- n1 /MarkedFraction

    # Likelihood
    # Binomial distribution for observed recaptures
```

```
    m2 ~ dbin(MarkedFraction, n2)
    }

    ",fill = TRUE)
sink()

# Bundle data
jags.data <- list("n1", "n2", "m2")

# Initial values.
jags.inits <- function(){ list(MarkedFraction=runif(1, min=0, max=1))}

model.file <- 'TwoSampleCR.txt'

# Parameters monitored
jags.params <- c("N.hat", "MarkedFraction")

# Call JAGS from R
jagsfit <- jags(data=jags.data, inits=jags.inits, jags.params,
                n.chains = 3, n.thin = 1, n.iter = 2000,
                n.burnin = 1000, model.file)
print(jagsfit, digits=3)
plot(jagsfit)

par(mfrow=c(3,1))
hist(jagsfit$BUGSoutput$sims.array[,,1], main="",
     xlab="Marked fraction")
hist(jagsfit$BUGSoutput$sims.array[,,2], main="",
     xlab="Population estimate")
plot(jagsfit$BUGSoutput$sims.array[,,1],
     jagsfit$BUGSoutput$sims.array[,,2], xlab="Marked fraction",
     ylab="Population estimate")
```

The last four lines of code illustrate that it can be instructive to examine not only the posterior distributions but also the underlying relationship(s); see bivariate plot in Figure 5.1. Here it is clear that N.hat is inversely related to the model parameter for marked fraction, but it may be less obvious in situations with more complex models.

If you were managing this fish population (e.g., a community lake), how might you use the estimate of abundance? Try a few runs (simulation and model fitting) to see whether this level of uncertainty would be tolerable for management. In doing so, keep in mind that this assumed capture probability (0.2) is quite high and would be difficult to achieve in many field situations. Is

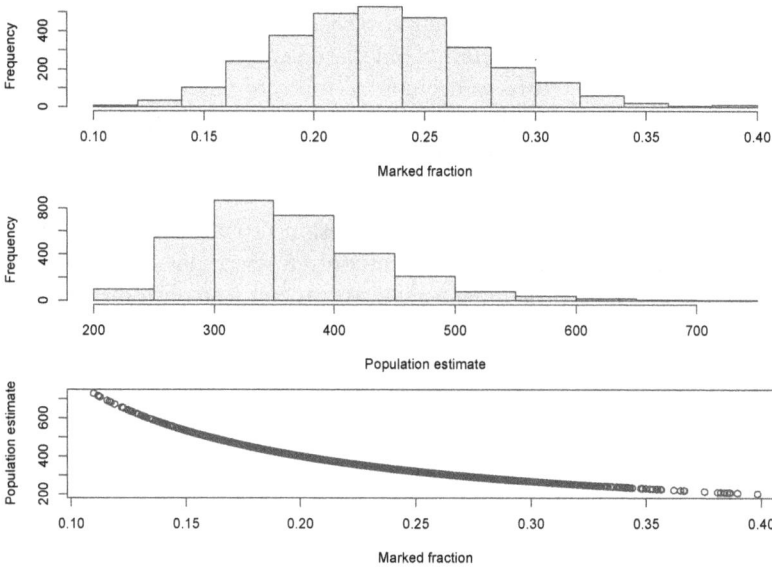

FIGURE 5.1 Estimated posterior distributions for marked fraction and population size and bivariate plot of posterior samples from two-sample mark-recapture study.

there a rough lower bound for capture probability, below which the cost and effort of doing a study would not be justified?

An example with empirical data is included as Appendix A.2.

5.3 Binomial-mixture

A population estimate can sometimes be obtained from count data when replicate survey samples are taken (Royle, 2004; Kéry and Schaub, 2012). One fisheries example is the use of side-scan sonar to count adult Atlantic sturgeon (*Acipenser oxyrinchus oxyrinchus*), which are large and distinctive enough to be identifiable on side-scan images (Flowers and Hightower, 2013, 2015). Sites were assumed to be closed (no movement in or out, no mortality) for the brief duration of the survey. Each site had an unknown number of sturgeon, but it was not assumed that every sturgeon was counted (detected). Counts could vary among survey passes due to a variety of factors, including a fish's orientation, overlap with other fish or structure, or distortion in the image. Thus the counting process was modeled using a binomial distribution, where "size" is the number of sturgeon present at the site, and "prob" is the probability of being detected on that survey pass. The detection probability is comparable to a capture probability but here sampling was unobtrusive. This type of model is referred to as a binomial-mixture model (or N-mixture model). It attempts to separate the biological parameter (site abundance) from the observational parameter (detection probability) (Royle, 2004; Kéry and Schaub, 2012).

Consider a site with 20 individuals and a detection probability of 0.5. In an actual field survey, there would typically be only a few replicate passes per site. Here we simulate a large sample of replicates to get a smooth distribution of counts:

```
rm(list=ls()) # Clear Environment

# Simulate experiment
N.true <- 20   # Number of individuals at site
p.true <- 0.5 # Detection probability
Replicates <- 1000
Counts <- rbinom(n=Replicates, size=N.true, prob=p.true)
  # Vector of replicate counts

par(mfrow=c(1,1)) # Ensure plot window is reset to default setting
hist(Counts, main="")
table(Counts)
```

The expected count is 10 (N.true * p.true); the histogram and table show the frequency of different values. One run produced minimum and maximum counts of 3 and 16. Counting 20 would be possible but extremely unlikely. One way to estimate empirically the probability of counting 15 or more is the following expression, which uses square brackets to subset the Counts vector: length(Counts[Counts>15])/Replicates. Try Counts[Counts>15] in the Console to make clear how this works. Experiment with different detection probabilities to find a level that produces occasional counts of 20 individuals.

To simulate a full field study, we need to add code for multiple sites. The number of visits per site can vary but, for simplicity, let's assume 30 sites and three replicate visits to each site. Kéry and Schaub (2012) recommended at least 10-20 sites and at least two observations per site. It is assumed that the number of sturgeon per site varies about an underlying mean, so having a large sample of sites will aid in estimating that mean. There are a variety of ways to model the variation in abundance among sites. Here we choose the simplest option of using a Poisson distribution, which has only one parameter to estimate:

```
rm(list=ls()) # Clear Environment

Sites <- 30
Replicates <- 3
lam.true <- 20 # Mean number per site
p.true <- 0.5 # Detection probability

Counts <- matrix(data=NA, nrow=Sites, ncol=Replicates)
Site.N <- rpois(n = Sites, lambda = lam.true) # True abundance per site
for (i in 1:Sites){
   Counts[i,] <- rbinom(n = Replicates, size = Site.N[i], prob = p.true)
   } #i
head(Site.N)
head(Counts)
```

The matrix() function creates an empty matrix (values "Not Available"), which will hold the replicate counts for each site. The simulation has a hierarchical structure. The number of sturgeon at each site is drawn from a Poisson distribution, using the shared mean lam.true and the rpois() function. We then loop over sites to generate the binomially-distributed replicate counts, given abundance at that site. We can use the head() function to look at the relationship between true abundance (Site.N) and counts for the first few sites. In one example run, the first site had 25 sturgeon and produced replicate counts of 11, 14, and 13. Try running rbinom(n=3, size=25, prob=x) multiple times in the Console with x=0.5 versus a probability near 0 or 1. The magnitude and variation among replicate counts within a site provide information about the detection probability and true site abundance.

```
# Load necessary library packages
library(rjags)
library(R2jags)

# Specify model in BUGS language
sink("N-mix.txt")
cat("
model {

# Priors
   lam.est ~ dunif(0, 100)  # uninformative prior for mean abundance
   p.est ~ dunif(0, 1)

# Likelihood
```

```
# Biological model for true abundance
for (i in 1:Sites) {
    N.est[i] ~ dpois(lam.est)
    # Observation model for replicated counts
    for (j in 1:Replicates) {
        Counts[i,j] ~ dbin(p.est, N.est[i])
        } # j
    } # i
totalN <- sum(N.est[])
    # Calculated variable: total population size across all sites
}
",fill = TRUE)
sink()

# Bundle data
jags.data <- list("Counts", "Sites", "Replicates")

# Initial values
Nst <- apply(Counts, MARGIN=1, max) + 1 # Kéry and Schaub
jags.inits <- function(){ list(lam.est=runif(n=1, min=0, max=100),
                               N.est=Nst, p.est=runif(1, min=0, max=1))}

model.file <- 'N-mix.txt'

# Parameters monitored
jags.params <- c("lam.est", "p.est",
                 #"N.est",   # To see individual site estimates
                 "totalN")

# Call JAGS from R
jagsfit <- jags(data=jags.data, jags.params, inits=jags.inits,
                n.chains = 3, n.thin=1, n.iter = 100000,
                model.file)
print(jagsfit)
par(mfrow=c(1,1)) # Reset plot panel
plot(jagsfit)

# Estimated posterior distribution for total population size
hist(jagsfit$BUGSoutput$sims.list$totalN,
     xlab="Estimated total pop size", main="")
abline(v=sum(Site.N), lty=3, lwd=3)
```

The JAGS code includes an uninformative prior for detection probability. The prior for mean site abundance is more subjective, but the upper bound could be based on the literature or knowledge about the system. As in the simulation code, the JAGS code has a hierarchical structure. Estimated overall mean abundance is used to draw site-specific abundance estimates from the shared Poisson distribution. The observations are binomially distributed, given the estimated abundance at each site. It is sometimes of interest to look at total abundance over sites, so the site estimates are summed (totalN).

It is important to provide initial values for site abundance. JAGS will terminate the run if a nonsensical event occurs (e.g., count greater than the current estimate of abundance at a site). Kéry and Schaub (2012) provide example code using the apply() function, which is a generic function for applying (hence the name) other functions. In this case, the max() function is applied to the Count matrix by row (argument MARGIN=1). The returned maximum value for each row is incremented by 1, implying that there is likely to be at least one more sturgeon present than the maximum observed count. The remaining parameters are initialized using settings that match the priors. The final lines of R code show the estimated posterior distribution for total abundance (summed over sites). The abline function is used to add to the plot the true total abundance (dotted vertical line).

Estimates are generally reliable for these assumed parameter values and study design settings. Convergence was sometimes slow, so a large number of updates was used. It is worthwhile to run the combined code (simulation plus analysis) several times to see variation in the estimates and their credible intervals. Median estimates of mean site abundance and detection probability tend to be close to the assumed values, although credible intervals are fairly wide. In sample runs, the mode of the posterior distribution for total abundance (Figure 5.2) was close to the true value, but the distribution was skewed and had a high upper bound for the credible interval.

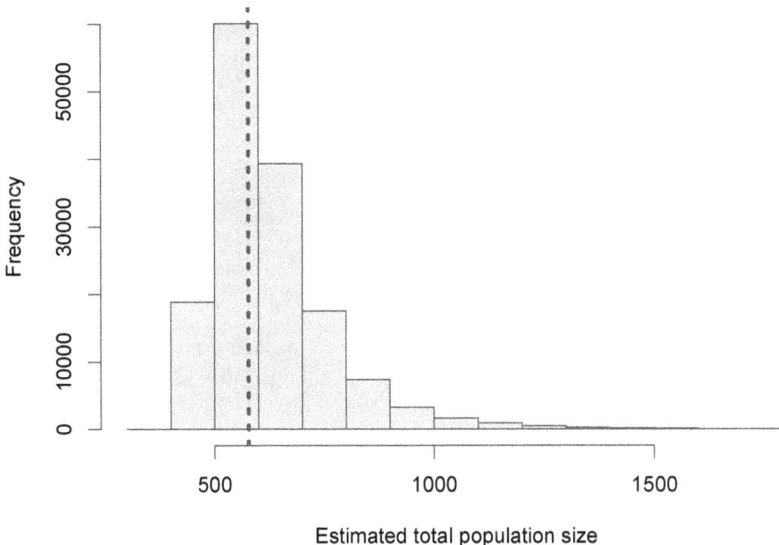

FIGURE 5.2 Estimated posterior distribution for total population size (summed over sites) and true total (dotted vertical line) based on binomial-mixture study.

This model is a good illustration of the potentially problematic nature of pV, the estimated number of effective parameters. A nominal count of parameters would be around 32, to account for detection probability, mean site abundance, and the site-specific abundance estimates using that mean. However, estimates from several runs were around 60. Kéry and Schaub (2012) note that hierachical models (such as this one) can be problematic for estimating the number of effective parameters (and therefore for DIC). We can look at pV for the various models in this book, to get some experience about when to trust pV and DIC estimates.

As usual, keep in mind that this is a very optimistic scenario. We are estimating mean abundance and detection probability using 30 sites that function as spatial replicates. The data are generated using a detection probability that is high and constant over sites. In an actual field study, there would be additional variation in detection probability among sites, and a binomial distribution would only be an approximation to the field sampling process that generates the observed counts. We have also assumed that variation among sites follows a Poisson distribution. Other distributions that allow for more variation among sites, such as a negative binomial, are possible and can be much more challenging to fit.

5.4   Exercises

1. For the removal experiment, create a table of estimated credible intervals for population size, using several replicate simulations and a range of values for capture probability. What are the lower limits below which estimates are judged not to be reliable?

2. For a two-sample mark-recapture experiment with a true population size of 1000, produce a table of credible intervals for N.hat for different assumed capture probabilities. What sampling intensity would produce an estimate with uncertainty close to the Robson and Regier target for management (+/- 25%)?

3. Modify the binomial-mixture example to compare credible intervals for estimated mean abundance (lam.est) using fewer sites but more replicates, preserving the same total number of observations (e.g., ten sites and nine replicate survey visits). Can you suggest potential advantages and disadvantages of using fewer sites?

6

Survival

Information about survival rate is valuable for managing exploited stocks or working to rebuild rare or declining populations. Survival is affected by two broad categories of threats: fishing and natural mortality. Fishing obviously refers to harvest, from commercial or recreational sectors, but it could include mortality associated with fish that are caught and released. Catch-and-release fishing primarily occurs in the recreational sector, but can occur in the commercial sector due to regulations or lack of markets (sometimes referred to as "discard"). Natural mortality is a generic term for mortality sources other than fishing. There is typically a strong negative correlation between natural mortality rate and body size, indicating the predominant role of predation (Lorenzen, 1996). Other sources of natural mortality include starvation, disease and senescence. It is important as a first step to estimate overall survival reliably, but management can be improved by partitioning total mortality into fishing and natural component rates. We begin with approaches for estimating overall survival, then in Chapter 7 consider tagging and telemetry methods that provide insights about sources of mortality.

6.1 Age-composition

The simplest approach for estimating survival is to examine the age-composition of a sample of fish (i.e., pattern of decline in number or proportion as age increases). Age would typically be determined through laboratory work using otoliths or other hard parts that show annual patterns (Campana, 2001). This approach is usually applied to adult fish that would be expected to have similar survival rates. We begin with simulation code for generating the age distribution:

```
rm(list=ls()) # Clear Environment

# Simulate age-composition sampling
```

```
MaxAge <- 10
S.true <- 0.7 # Annual survival probability

# Simulated age vector at equilibrium (expected values)
N <- array(data=NA, dim=MaxAge)
N[1] <- 1000
for (j in 2:(MaxAge)){
  N[j] <- N[j-1] * S.true
} #j
age.vec <- seq(1,MaxAge, by=1)   # Sequence function: plot x axis
barplot(N, names.arg=age.vec, xlab="Age", ylab="Number")

SampleSize <- 30
AgeMatrix <- rmultinom(n=1, size=SampleSize, prob=N)
# Note: prob vector rescaled internally
AgeSample <- as.vector(t(AgeMatrix))
   # Transpose matrix then convert to vector
barplot(AgeSample, names.arg=age.vec, xlab="Age", ylab="Frequency")
```

The simulation tracks fish up to age 10 (maximum age tracked, not lifes-pan) with an arbitrary annual survival rate of 0.7. The looping code for age-composition starts with 1000 fish at age 1 and reduces that number at each successive age by the annual survival rate. This creates an equilibrium age vector (up to the maximum age), based on assumptions of constant recruitment (1000 each year) and survival. For example, the 700 age-2 fish would have been the survivors from the previous year's 1000 age-1 recruits. The initial number is arbitrary; the shape of the distribution and age proportions would be the same if we began with 1 or 1000 fish. The next section of code uses the equilibrium age vector as proportions and draws a sample of size 30 using a multinomial distribution (Section 3.1.3). To build some intuition about age sampling, try running the simulation code multiple times with different values for sample size, maximum age, and survival rate. As usual, keep in mind that this is a very optimistic scenario, with constant recruitment and survival and unbiased survey sampling. The only variation reflected in these samples is due to sampling from the multinomial distribution (as is evident if a large age sample is drawn).

The JAGS code for analysis is straightforward and mirrors the simulation code. Note that the equilibrium age vector used as proportions in the dmulti() function must be rescaled manually to sum to 1. Convergence is fast and the estimate of survival rate is moderately precise, despite the small (but perfectly unbiased) sample. For example, the 95% credible interval for one run was 0.57-0.80. The final two lines of R code show the posterior distribution for survival rate, compared to the true value.

```
# Load necessary library packages
library(rjags)
library(R2jags)

# JAGS code
sink("AgeComp.txt")
cat("
model{
    # Priors
    S.est ~ dunif(0,1)  # Constant survival over sampled age range

    N[1] <- 1000
    # Generate equilibrium age proportions
    for (j in 2:MaxAge){
      N[j] <- N[j-1]*S.est
    } # j
    N.sum <- sum(N[]) # Rescale so that proportions sum to 1
    for (j in 1:MaxAge){
    p[j] <- N[j]/N.sum
    } #j
    # Likelihood
    AgeSample[] ~ dmulti(p[], SampleSize)
    # Fit multinomial distribution based on constant survival
}
    ",fill=TRUE)
sink()

# Bundle data
jags.data <- list("AgeSample", "SampleSize", "MaxAge")

# Initial values
jags.inits <- function(){ list(S.est=runif(n=1, min=0, max=1))}

model.file <- 'AgeComp.txt'

# Parameters monitored
jags.params <- c("S.est")

# Call JAGS from R
jagsfit <- jags(data=jags.data, jags.params, inits=jags.inits,
                n.chains = 3, n.thin=1, n.iter = 4000, n.burnin=2000,
                model.file)
print(jagsfit)
plot(jagsfit)
```

```
# Look at posterior distribution versus true value
hist(jagsfit$BUGSoutput$sims.list$S.est,
     xlab="Estimated survival rate", main="")
abline(v=S.true, lty=3, lwd=3)
```

Note that we are not estimating the age proportions as parameters but rather the single parameter for survival rate (which is used to predict the equilibrium age proportions). Thus the number of effective parameters (pV) should be close to one. A few lines of code can be added to compare the predicted and observed age distributions:

```
# Add code to compare observed and predicted age comp
N.hat <- array(data=NA, dim=MaxAge)
N.hat[1] <- 1000
for (j in 2:(MaxAge)){
  N.hat[j] <- N.hat[j-1] * jagsfit$BUGSoutput$median$S.est
} #j
par(mfrow=c(1,2))
barplot(AgeSample, names.arg=age.vec, xlab="Age",
        ylab="Number", main="Observed")
barplot(N.hat, names.arg=age.vec, xlab="Age",
        ylab="Number", main="Predicted")
par(mfrow=c(1,1)) # Reset plot window
```

The predicted age distribution looks like a smoothed version of the observed, which makes sense given that we are estimating average survival over the sampled range of ages. When a poor estimate of the survival rate occurs, you will likely notice an odd pattern in the simulated age-composition sample. Try running the code several times to look at variation in age-composition and estimated survival. Increasing the sample size will bring the observed pattern closer to the equilibrium age-composition.

Thus far, we have used small sample sizes for these simulated studies. They are sufficient here, because the simulation only includes variation due to sampling from the multinomial distribution. It is also convenient to work with a small sample size when studying the code and simulation data. Sample sizes are sometimes small in field studies, and it is important to establish that a study design will provide reliable results when using an achievable sample size. But it can also be informative to test the analytical approach with a very large sample size. This provides a best-case scenario for study results and ensures that estimates approach the true values (are essentially unbiased). For example, three replicate median estimates of the survival rate were 0.71, 0.69, and 0.70 for a sample size of 2000 fish.

An example using fishery catch data and a catch curve analysis is included as Appendix A.3.

6.1.1 Accounting for gear selectivity

The traditional approach for estimating survival from age-composition data was known as catch curve analysis. It could be done using linear regression (thus doable by hand in the pre-computer era) by log-transforming the age sample (Ricker, 1975). Gear selectivity was addressed subjectively, by viewing a plot of the log-transformed catches. Ricker (1975) described the catch curve pattern as having ascending and descending limbs, with the ascending limb assumed to reflect selectivity and survival. The descending limb was assumed to reflect only survival (age range of "fully selected" fish). The recommended approach was to begin the catch curve analysis at the age of peak catch or the next older age, which was assumed to limit the analysis to fully selected fish. For a catch curve analysis or the approach given here, declining selectivity for older fish would cause the survival estimator to be biased.

Given sufficient data, it is possible to use age-composition data to estimate jointly the survival rate and gear selectivity. This makes it possible to use the full range of ages and avoids making a subjective decision about where to start the analysis. A practical example might be a trawl survey where vulnerability to the gear increases with age to an asymptote of 1 (fully selected). This pattern in gear selectivity can be described using a logistic function:

$$S_a = \frac{1}{1+e^{k*(a-a_{0.50})}}$$

Open a new R Script window and run the following lines of code to study the logistic gear selectivity pattern. Default values for the slope (k) and age at 50% selectivity ($a_{0.50}$) are arbitrary, but chosen to increase to a well-defined asymptote before age 10. Try different values for the slope and $a_{0.50}$ to understand how the two parameters affect the shape of the curve. Note that any analysis of gear selectivity will only be successful if all ages share the survival rate, and there is a sufficient age range to fully characterize the asymptote. For example, imagine the outcome of an analysis using data only for ages 1-5.

```
rm(list=ls()) # Clear Environment
MaxAge <- 10
# Add in selectivity
k <- 2 # Slope for logistic function
a_50 <- 4 # Age at 0.5 selectivity
```

```
# Age selectivity
Sel <- array(data=NA, dim=MaxAge)
for (j in 1:(MaxAge)){
  Sel[j] <- 1/(1+exp(-k*(j-a_50)))
} #j
age.vec <- seq(1,MaxAge, by=1)  # Sequence function: plot x axis
par(mfrow=c(1,1))
plot(age.vec, Sel, xlab="Age", ylab="Gear selectivity")
```

Now we can return to the original R Script window, replacing the original code with the following simulation code, extended to include gear selectivity:

```
rm(list=ls()) # Clear Environment

# Simulate age-composition sampling with age-specific selectivity
MaxAge <- 10
S.true <- 0.7 # Annual survival probability

# Parameters for selectivity curve
k <- 2 # Slope for logistic function
a_50 <- 4 # Age at 0.5 selectivity

# Age selectivity
Sel <- array(data=NA, dim=MaxAge)
for (j in 1:(MaxAge)){
  Sel[j] <- 1/(1+exp(-k*(j-a_50)))
} #j
age.vec <- seq(1,MaxAge, by=1)  # Sequence function: plot x axis values
par(mfrow=c(1,1))
plot(age.vec, Sel, xlab="Age", ylab="Gear selectivity")

# Simulated age vector, including gear selectivity
N <- array(data=NA, dim=MaxAge)
N[1] <- 1000
for (j in 2:(MaxAge)){
  N[j] <- N[j-1] * S.true
} #j
N_Sel <- N * Sel
barplot(N_Sel, names.arg=age.vec, xlab="Age", ylab="Number",
        main="Expected age distribution")

SampleSize <- 100
AgeMatrix <- rmultinom(n=1, size=SampleSize, prob=N_Sel)
  # Note: prob vector rescaled internally
```

```
AgeSample <- as.vector(t(AgeMatrix))
  # Transpose matrix then convert to vector
barplot(AgeSample, names.arg=age.vec, xlab="Age",
        ylab="Frequency", main="Sample age distribution")
```

The sample age distribution now reflects both gear selectivity and survival. We use a relatively large sample size and test whether it is possible to estimate all three parameters (survival rate and two parameters for selectivity):

```
# JAGS analysis
library(rjags)
library(R2jags)

sink("AgeComp.txt")
cat("
model{
    # Priors
    S.est ~ dunif(0,1)  # Constant survival over sampled age range
    a50.est ~ dunif(0, MaxAge)
    k.est ~ dlnorm(0, 1E-6) # Slope > 0 for increasing selectivity

    N[1] <- 1000
    # Generate equilibrium age proportions
    for (j in 2:MaxAge){
      N[j] <- N[j-1]*S.est
    } # j
    for (j in 1:MaxAge){
    Sel[j] <- 1/(1+exp(-k.est*(j-a50.est)))
      N.Sel[j] <- N[j] * Sel[j]
    }
    N.sum <- sum(N.Sel[]) # Rescale so that proportions sum to 1
    for (j in 1:MaxAge){
    p[j] <- N.Sel[j]/N.sum
    } #j
    # Likelihood
    AgeSample[] ~ dmulti(p[], SampleSize)
    # Fit multinomial distribution based on constant survival
}
    ",fill=TRUE)
sink()

# Bundle data
```

```
jags.data <- list("AgeSample", "SampleSize", "MaxAge")

# Initial values
jags.inits <- function(){ list(S.est=runif(n=1, min=0, max=1),
                               a50.est=runif(n=1, min=0, max=MaxAge),
                               k.est=rlnorm(n=1, meanlog=0, sdlog=1))}
                               # Arbitrary low initial values for k.est

model.file <- 'AgeComp.txt'

# Parameters monitored
jags.params <- c("S.est", "a50.est", "k.est")

# Call JAGS from R
jagsfit <- jags(data=jags.data, jags.params, inits=jags.inits,
                n.chains = 3, n.thin=1, n.iter = 20000,
                model.file)
print(jagsfit)
plot(jagsfit)
```

The age at 50% selectivity should be less than the maximum age, so a uniform prior is appropriate. The slope of the selectivity curve does not have an obvious upper bound, so following McCarthy (2007), we use an uninformative lognormal prior. Restricting the slope to positive values is consistent with a survey gear that underrepresents smaller fish. Convergence can be slow when using a lognormal prior, so check Rhat values and increase the number of updates (n.iter) if needed.

There are several reasons why this analysis works (at least for these assumed parameter values). We have an unbiased sample of size 100 from the expected age distribution (which reflects both survival and selectivity). The most important reason is that survival rate is constant over the entire age range. Gear selectivity and survival rate are confounded for young fish but not for older fish that are fully vulnerable to the survey. Thus the shape of the gear-selectivity function matters, as does the age range included in the analysis. It would not be possible to jointly estimate gear selectivity and survival rate if selectivity increased across the surveyed age range. And as before, the analysis is a best-case scenario in that the only variation is due to sampling from the multinomial distribution. Gear selectivity varies by age, but all ages share the same selectivity parameters.

One way to improve on this analysis would be to include auxiliary data on selectivity. We could accomplish this by conducting a short-term tagging study with (for simplicity) equal numbers tagged by age. Having a tagged subpopulation with a known age structure provides direct information on gear selectivity, unlike the age-composition data where selectivity and survival rate are partially confounded. The logistical challenge in adding this auxiliary study is in obtaining sufficient tag returns through survey sampling (the gear providing the age-composition data).

The following version of the code includes auxiliary tag-return data:

```r
rm(list=ls()) # Clear Environment

# Simulate age-composition sampling with age-specific selectivity
MaxAge <- 10
S.true <- 0.7 # Annual survival probability

# Parameters for selectivity curve
k <- 2 # Slope for logistic function
a_50 <- 4 # Age at 0.5 selectivity

# Age selectivity
Sel <- array(data=NA, dim=MaxAge)
for (j in 1:(MaxAge)){
  Sel[j] <- 1/(1+exp(-k*(j-a_50)))
} #j
age.vec <- seq(1,MaxAge, by=1)  # Sequence function: x-axis labels
par(mfrow=c(1,1))
plot(age.vec, Sel, xlab="Age", ylab="Gear selectivity")

# Simulated age vector, including gear selectivity
N <- array(data=NA, dim=MaxAge)
N[1] <- 1000
for (j in 2:(MaxAge)){
  N[j] <- N[j-1] * S.true
} #j
N_Sel <- N * Sel
barplot(N_Sel, names.arg=age.vec, xlab="Age", ylab="Number",
        main="Expected age distribution")

SampleSize <- 100
AgeMatrix <- rmultinom(n=1, size=SampleSize, prob=N_Sel)
  # prob vector is rescaled internally
AgeSample <- as.vector(t(AgeMatrix))
  # Transpose matrix then convert to vector
barplot(AgeSample, names.arg=age.vec, xlab="Age", ylab="Frequency",
        main="Sample age distribution")
```

```
# Auxiliary tag-return data
TagReturns <- 30
RetMatrix <- rmultinom(n=1, size=TagReturns, prob=Sel)
  # prob vector is rescaled internally
RetVector <- as.vector(t(RetMatrix))
  # Transpose matrix then convert to vector
barplot(RetVector, names.arg=age.vec, xlab="Age", ylab="Frequency",
        main="Tag returns")

# JAGS analysis
library(rjags)
library(R2jags)

sink("AgeComp.txt")
cat("
model{
    # Priors
    S.est ~ dunif(0,1)  # Constant survival over sampled age range
    a50.est ~ dunif(0, MaxAge)
    k.est ~ dlnorm(0, 1E-6) # Slope > 0 for increasing selectivity

    N[1] <- 1000
    # Generate equilibrium age proportions
    for (j in 2:MaxAge){
      N[j] <- N[j-1]*S.est
    } # j
    for (j in 1:MaxAge){
    Sel[j] <- 1/(1+exp(-k.est*(j-a50.est)))
     N.Sel[j] <- N[j] * Sel[j]
     }
    N.sum <- sum(N.Sel[]) # Rescale so that proportions sum to 1
    for (j in 1:MaxAge){
    p[j] <- N.Sel[j]/N.sum
    } #j
    # Likelihood
    AgeSample[] ~ dmulti(p[], SampleSize)
      # Fit multinomial distribution based on constant survival

    # Additional likelihood for tag-return data
    Sel.sum <- sum(Sel[])
    for (j in 1:MaxAge){
    Sel.p[j] <- Sel[j]/Sel.sum
    } #j
    RetVector[] ~ dmulti(Sel.p[], TagReturns)
```

```
}
    ",fill=TRUE)
sink()

# Bundle data
jags.data <- list("AgeSample", "SampleSize", "MaxAge",
                  "TagReturns", "RetVector")

# Initial values
jags.inits <- function(){ list(S.est=runif(n=1, min=0, max=1),
                        a50.est=runif(n=1, min=0, max=MaxAge),
                        k.est=rlnorm(n=1, meanlog=0, sdlog=1))}
                   # Arbitrary low initial value for k.est


model.file <- 'AgeComp.txt'

# Parameters monitored
jags.params <- c("S.est", "a50.est", "k.est")

# Call JAGS from R
jagsfit <- jags(data=jags.data, jags.params, inits=jags.inits,
                n.chains = 3, n.thin=1, n.iter = 20000,
                model.file)
print(jagsfit)
plot(jagsfit)
```

There are three changes to the code in this version. First, the vector of random tag returns is generated just below the random survey catch vector. We assume a sample size of 30 returned tags, allocated randomly among ages using a multinomial distribution and the survey gear selectivity pattern. The returns by age are a direct measure of gear selectivity, because of our assumption that equal numbers were tagged by age. You can examine the barplot for the tag returns to get a sense of what the auxiliary study contributes to the overall analysis. The next code change is adding a second likelihood component to make use of the tag-return data. The third change is to pass these additional data (tag return total and by age) to JAGS.

Having direct information on survey gear selectivity can improve estimates of survival and selectivity, depending on the sample sizes for age-composition and returned tags. And more importantly, this modification illustrates how auxiliary data can be added in JAGS to refine an analysis. Writing code to fit the specifics of a field study is valuable for planning and analysis, and it is a major advantage over using canned software. The remaining chapters will include other examples where multiple data sources can be combined to improve estimates.

6.1.2   Cohort model

One limitation of the above analysis (Section 6.1) or a traditional catch curve analysis is that recruitment is assumed to be constant (or can be averaged over). Catch curve analyses are generally robust to this assumption (Allen, 1997), but year-class variation can sometimes be quite extreme. The following approach addresses that concern by modeling each year-class (or cohort) separately. The approach is similar to a virtual population analysis (VPA) or cohort analysis (Ricker, 1975) in that a catch matrix provides age- and year-specific information that is used to track cohorts across years. A VPA is traditionally done using harvest data, but here, we assume the catch-at-age matrix contains survey data. Survey information allows for modeling relative abundance across time, whereas a traditional VPA is used to estimate absolute abundance.

The following simulation code provides the survey catch-at-age matrix. The number of ages and years is arbitrary, but we are estimating survival rate for each year, so there should be enough ages to provide a decent sample size. The annual survival rate is drawn from a uniform distribution between arbitrary lower and upper bounds. Having a fair degree of annual variation in survival rate makes for a better test of our ability to estimate those values.

```
rm(list=ls()) # Clear Environment

# Simulation to create survey catch-at-age matrix
# Assume ages included in matrix are fully selected
A <- 4 # Arbitrary number of ages in survey catch matrix
Y <- 10
LowS <- 0.2 # Arbitrary lower bound
HighS <- 0.9 # Arbitrary upper bound
MeanS <- (LowS+HighS)/2
AnnualS <- runif((Y-1), min=LowS, max=HighS)

MeanR <- 40000 # Arbitrary mean initial cohort size (relative)
N <- array(data=NA, dim=c(A, Y))
N[1,] <- rpois(Y, MeanR) # Starting size of each cohort at first age
for (a in 2:A){
  N[a,1] <- rpois(1,MeanS^(a-1)*MeanR) # Starting sizes ages 2+, year 1
    for (y in 2:Y){
      N[a,y] <- rbinom(1, N[a-1, y-1], AnnualS[y-1]) # Fill in older ages
```

```
    } #y
  } #a

SurveyCapProb <- 0.001 # Capture probability
SurveyC <- array(data=NA, dim=c(A, Y))
# Generate survey catches
for (a in 1:A){
  for (y in 1:Y){
    SurveyC[a,y] <- rbinom(1, N[a, y], SurveyCapProb)
  } #y
} #a
```

The model tracks coded ages 1:A, which could represent a range of older ages
with comparable survival rates. True (relative) abundance at age 1 (N[1,]) is
drawn from a Poisson distribution, based on an arbitrary mean initial cohort
size. Initial numbers for ages 2:A in the first year are also obtained from a Poisson distribution, where the expected number at age a is obtained by a product
of survival rates multiplied by mean recruitment. For example, the expected
number is MeanR * S at age 2 (adjusting for survival over one year), MeanR
* S^2 for age 3, etc. The code then loops over years (columns) and ages (rows)
to fill in the rest of the matrix, using a binomial distribution to simulate the
survival process (and using the year-specific survival rate). The survey catch-
at-age matrix includes variation due to sampling (binomally-distributed using
an assumed capture probability of 0.001), as well as year-class strength and survival. Click on N and SurveyC in the Environment window to compare the two
matrices in RStudio Source windows. The assumed capture probability can be
adjusted relative to MeanR to provide adequate (or inadequate) survey catches.

The JAGS code for estimating cohort size and annual survival rate is generally
similar to the simulation code. Prior distributions and initial values for predicted
survey catches use lognormal distributions. Using high initial values for the
abundance parameters reduces the risk of a JAGS error (e.g., estimated survey
"abundance" at time y-1 less than observed survey catch at time y). The
likelihood for the survey catch matrix uses a Poisson distribution.

```
# Load necessary library packages
library(rjags)    # Package for fitting JAGS models from within R
library(R2jags)   # Package for fitting JAGS models. Requires rjags

# JAGS code
sink("CatchAge.txt")
cat("
model {
```

```
    # Priors
    for (y in 1:Y) {
      Recruits[y] ~ dlnorm(0, 1E-6) # Non-negative values
      SurveyC.est[1,y] <- trunc(Recruits[y]) }

    for (a in 2:A){
      Initials[a-1] ~ dlnorm(0, 1E-6)
      # Non-negative; offset index to run vector from 1 to A-1
      SurveyC.est[a,1] <- trunc(Initials[a-1]) }

  meanS.est <- prod(S.est)^(1/(Y-1))
  for (y in 1:(Y-1)) {
    S.est[y] ~ dunif(0,1)
    } #y

    # Population projection
    for (a in 2:A) {
    for (y in 2:Y) {
      SurveyC.est[a,y] ~ dbin(S.est[y-1], SurveyC.est[a-1,y-1])
      } } # a, y

    #Likelihood
    for (a in 1:A) {
    for (y in 1:Y) {
      SurveyC[a,y]~dpois(SurveyC.est[a,y])
      } } #a #y
}
    ",fill=TRUE)
sink()

# Bundle data
jags.data <- list("Y", "A", "SurveyC")

# Initial values
jags.inits <- function(){ list(Recruits=rlnorm(n=Y,
                                                meanlog=10, sdlog=1),
                          Initials=rlnorm(n=(A-1),
                                                meanlog=10, sdlog=1))}

model.file <- 'CatchAge.txt'

# Parameters monitored
jags.params <- c("S.est", "meanS.est")
```

```
# Call JAGS from R
jagsfit <- jags(data=jags.data, jags.params, inits=jags.inits,
                n.chains = 3, n.thin=1, n.iter = 40000,
                model.file)
print(jagsfit)
plot(jagsfit)

plot(seq(1:(Y-1)), jagsfit$BUGSoutput$median$S.est, ylab="Survival",
     xlab="Year", type="b", pch=19, ylim=c(0,1))
points(AnnualS, type="b", lty=3)
```

The final three lines of code compare the annual survival estimates and true values (Figure 6.1). The plotted values are medians, which provide a better measure of central tendency for skewed posterior distributions (Kruschke, 2021). The plot option "type=b" produces a line of connected dots (i.e., *both* points and line segments), and the "pch=19" option selects for a filled dot. An online search for "r plot options pch" will provide the various choices for plotting symbols. The range for the y-axis is fixed at 0-1 to ensure that the plot will contain the full range of both the estimates and the true values. The points() function adds to the plot the true values.

The model provides reliable estimates of annual (and mean) survival, although credible intervals for annual survival tend to be wide. Run the simulation and

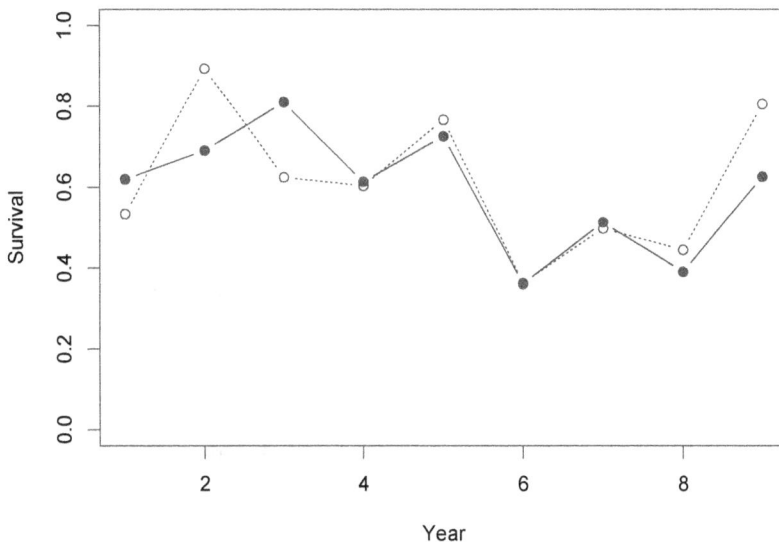

FIGURE 6.1 Example run showing true (dotted, open) and estimated (solid, filled) annual survival rate from a cohort-based analysis of survey catch-at-age data.

model-fitting code multiple times to observe model performance. It would be worthwhile to test the approach using more variability in initial cohort sizes; for example, by using a negative binomial distribution rather than a Poisson.

The ability to capture the temporal pattern in survival depends on the capture probability, relative to MeanR. Survival rates varied randomly in this simulation, but other temporal patterns might be of interest in a real-world analysis. For example, a biologist might want to look for a response to a regulation change or an environmental event. If annual survival is thought simply to vary randomly, a model using a hierarchical structure may be preferred. In that case, annual survival estimates would be drawn from a common distribution with a shared mean survival rate.

It is important to keep in mind that the predicted survey catches are not estimates of true abundance but simply scaled to the survey catches. Millar and Meyer (2000) illustrate a Bayesian approach for estimating true abundance from survey *and* harvest catch-at-age matrices.

6.2   Telemetry-based

The focus for this section will be on estimating survival rate using fixed receiver stations and telemetry tags such as transmitters or passive integrated transponder (PIT) tags. Letting fish "detect themselves" eliminates the need for costly field surveys that often result in low detection probabilities. For example, Hewitt et al. (2010) showed that survival estimates using fixed stations and automated detections were much better than those obtained using only survey recaptures. Similarly, Hightower et al. (2015) estimated survival rate of sonic-tagged Atlantic sturgeon which were detected as they migrated past receivers located in estuaries and rivers. The key in studies of this type is either to have many monitoring sites or to position sites in areas where tagged fish are likely to pass.

These studies produce detections that can be organized as a capture history matrix. Each row of the matrix is a tagged fish, and each column is a time period. The period can be chosen based on whatever time scale is relevant and practical; for example, monthly or quarterly detections would provide insights about seasonal survival. Each matrix cell indicates whether a fish was detected (1) or not (0) in that period. A fish not detected could be dead or simply did not move within range of a receiver during that period. Interpreting the capture history for an individual depends on the pattern. There is no ambiguity about an individual with a capture history of 11011. It was alive to the end of the study but was not detected on the third occasion. A sequence such as 11000 is not as straightforward because its fate is uncertain for the final three occasions. A model is essential for making inferences in such cases.

The capture history data can be analyzed using a Cormack-Jolly-Seber (CJS) model (Royle, 2008; Kéry and Schaub, 2012). The CJS model is based only on captures or detections of tagged fish, which are assumed to be representative of the population as a whole. Technically, the model provides an estimate of apparent survival, because we are not able to distinguish between mortality and permanent emigration. If the study area is closed to migration, then the estimate can be interpreted as a true survival rate.

Our simulation is for the most basic case, with survival and detection probabilities that are constant over time. Extended models that allow for time variation are presented by Kéry and Schaub (2012). The number of occasions and sample size are arbitrary but they serve as useful replicates here because the underlying parameters are constant over time and individuals. We assume a moderate survival rate (0.5) and a high detection probability (0.8). Detection probabilities of that magnitude are difficult to achieve by survey sampling for tagged fish but can be achieved using telemetry with an automated monitoring system (Hewitt et al., 2010).

```
rm(list=ls()) # Clear Environment

# Modified from Kéry and Schaub 2012, section 7.3
# Define parameter values
n.occasions <- 5            # Number of occasions
marked <- 50                # Number marked (all at time 1)
S.true <- 0.5
p.true <- 0.8

# Define function to simulate a capture-history (CH) matrix
simul.cjs <- function(S.true, p.true, marked, n.occasions){
  Latent <- CH <- matrix(0, ncol = n.occasions, nrow = marked)
  # Fill the CH matrix
  for (i in 1:marked){
    Latent[i,1] <- CH[i,1] <- 1    # Known alive state at release
    for (t in 2:n.occasions){
      # Bernoulli trial: does individual survive occasion?
      Latent[i,t] <- rbinom(1, 1, (Latent[i,(t-1)]*S.true))
      # Bernoulli trial: is individual recaptured?
      CH[i,t] <- rbinom(1, 1, (Latent[i,t]*p.true))
    } #t
  } #i
  return(CH)
}

# Execute function
CH <- simul.cjs(S.true, p.true, marked, n.occasions)
```

Our code, modified from Kéry and Schaub (2012), introduces the creation of a user-defined function, which is a way of setting off a discrete block of code that has a specific purpose. We pass to the function the necessary arguments (S.true, p.true, marked, n.occasions) and the function returns the desired result (capture-history (CH) matrix). The latent (hidden true) state and the CH observation are modeled using Bernoulli (Section 3.1.1) trials. The latent state at time 't' is based on the survival rate between times t-1 and t, but multiplied by the true state at time t-1. That results in live fish having a survival probability of 1 * S.true, whereas the probability for dead fish is 0 * S.true (ensuring that dead fish remain dead). A similar trick is used to generate the observations, given the known true states. Detection probability is 1 * p.true for live fish and 0 * p.true for dead fish. The matrix for latent state is internal to the function, but (for instructional purposes) it can be compared to the observed matrix by selecting and running lines 2-12 of the function. You should see the connection between the true latent state and observations (e.g., a true sequence of 11100 versus an observed sequence of 10100).

```r
# Load necessary library packages
library(rjags)
library(R2jags)

# Specify model in JAGS
sink("CJS_StateSpace.txt")
cat("
model {

# Priors and constraints
S.est ~ dunif(0, 1)
p.est ~ dunif(0, 1)

# Likelihood
for (i in 1:marked){
   z[i,1] <- 1 # Latent state 1 (alive) at time of release (time 1 here)
   for (t in 2:n.occasions){
      # State process. If z[t-1]=1, prob=S. If z=0, prob=0
      z[i,t] ~ dbern(z[i,t-1]*S.est)

      # Observation process. If z[t]=1, prob=p. If z=0, prob=0
      CH[i,t] ~ dbern(z[i,t]*p.est)
      } #t
   } #i
}
",fill = TRUE)
sink()

# Bundle data
jags.data <- list(CH = CH, marked = marked, n.occasions = n.occasions)
```

```
# Initial values for latent state z
init.z <- function(ch){
  state <- ch
  for (i in 1:dim(ch)[1]){
    n1 <- min(which(ch[i,]==1)) # Column with first detection (col 1)
    n2 <- max(which(ch[i,]==1)) # Column with last detection
    # Fish are known to be alive between n1 and n2
    state[i,n1:n2] <- 1
    state[i,n1] <- NA # Initial value not needed for release period
  }
  state[state==0] <- NA
  return(state)
}


jags.inits <- function(){list(S.est = runif(1, 0, 1),
                              p.est = runif(1, 0, 1), z = init.z(CH))}

# Parameters monitored
jags.params <- c("S.est", "p.est")

model.file <- 'CJS_StateSpace.txt'

jagsfit <- jags(data=jags.data, jags.params, inits=jags.inits,
            n.chains = 3, n.thin=1, n.iter = 4000, n.burnin=2000,
            model.file)
print(jagsfit)
plot(jagsfit)
```

The JAGS code for fitting the model uses our standard uninformative priors for the two probabilities (survival, detection probability). The likelihood mirrors the simulation code. The code is structured as a state-space model (Kéry and Schaub, 2012), with matrix 'z' representing the latent state, which is sometimes known but inferred in situations where the pattern is ambiguous. The observation process (detected or not) is modeled next, and depends on the biological process (latent state).

JAGS requires good initial values, or will terminate execution if it encounters an initial value that is inconsistent with the observations, such as detection of a fish that the model considers to be dead. To avoid that, we use the Kéry and Schaub (2012) approach of initializing the latent state to 1 between the first and last detections of an individual (user-defined function init.z()). Experiment in the Console with the which() function to see how the initialization function init.z works (e.g., which(CH[3,]==1)). It can also be helpful to inspect the output from init.z by entering z.test = init.z(CH) in the Console.

Estimates appear to be very reliable for this scenario, where capture histories for 50 fish are used to estimate only two time-independent parameters. Would you obtain usefully precise results with a sample size of 10 or 25 fish? How does reliability change as you reduce the assumed detection probability?

One important advantage of this telemetry method is that estimates are essentially produced in real time and on any time scale. For example, the method could be used to estimate average annual survival over the first few years after a regulation change. An extended version using annual survival rates (Kéry and Schaub, 2012) could be fitted to observations over a block of years with a management action taken midway.

6.3   Tag-based

Survival can be estimated by a multi-period study using tag returns. An important advantage of this approach is that field effort is limited to the tagging process because the fishery generates the tag returns. The design requires a release of tagged fish at the start of each period. The releases need not be the same, but here we assume a constant number released for simplicity. Like the telemetry method (Section 6.2), estimates can be produced for whatever time interval is relevant (e.g., monthly, yearly).

This study design is based on band-recovery models used in wildlife studies (Brownie et al., 1985; Williams et al., 2002). Those methods were pioneered for migratory birds, with bands returned from individuals that were harvested. In fisheries tagging studies, tags may be also reported from fish that are caught and released, but here we assume that returned tags reflect only harvest. Catch-and-release complicates the analysis because the tag "dies" but the fish is not killed. Jiang et al. (2007) provide analytical methods for addressing the combination of harvest and catch-and-release.

We begin by setting up matrices for tags-at-risk (tagged fish still alive) and the outcomes from each release. The matrix of outcomes or fates is analyzed row-by-row using a multinomial distribution (Section 3.1.3). For example, a fish tagged in period 1 could be caught in any period up to the end of the study. It could also end up in the final column for fish not seen again, which would account for fish that were still alive or died of natural causes or were harvested, but the tag was not reported. We do not assume here any knowledge of the tag-reporting rate. If that information were available (e.g., through use of high-reward tags that provide an assumed reporting rate of 100%), then it would be possible to partition total mortality into fishing and natural component rates (see Section 7.3). Here we keep things

simple and estimate only the tag-recovery rate, which is the product of the
exploitation rate and tag-reporting rate. In planning a tag-return study, the
best results will be obtained by ensuring a tag-reporting rate very close to
100%. Non-reporting results in lost information (a reduced sample size) and
poorer precision.

```
rm(list=ls()) # Clear Environment

# Parameter values for three-yr experiment
Periods <- 3  # Equal number of release and return periods assumed
NTags <- 1000 # Number released each period
S.true <- 0.8 # Survival rate
r.true <- 0.1 # Tag recovery rate

# Calculated value(s), array creation
TagsAtRisk <- array(data=NA, dim=c(Periods, Periods))
TagFate <- array(data=NA, dim=c(Periods, Periods+1))
# Extra column for tags not seen again

for (i in 1:Periods){
# Expected values for tags-at-risk, returned/not-seen-again
    TagsAtRisk[i,i] <- NTags
    TagFate[i,i] <-  TagsAtRisk[i,i] * r.true # Release period returns
  j <- i+1
  while(j <= Periods) {
      TagsAtRisk[i,j] <- TagsAtRisk[i,(j-1)]*S.true
      TagFate[i,j] <- TagsAtRisk[i,j] * r.true
      j <- j+1
    } #while
  TagFate[i, Periods+1] <- NTags-sum(TagFate[i,i:Periods])
     # Not seen again
  } #i

#Random trial (tags returned by period or not seen again)
RandRet <- array(data=NA, dim=c(Periods, Periods+1))
  for (i in 1:Periods) # Create upper diagonal matrix of random outcomes
  {
    RandRet[i,i:(Periods+1)] <- t(rmultinom(1, NTags,
                                     TagFate[i,i:(Periods+1)]))
  } #i
```

Tags at risk are contained in an upper triangular matrix, with the diagonal
element being the number released (NTags) and subsequent elements obtained
as the previous number at risk multiplied by the survival rate. We arbitrarily
assume Periods=3, a true survival rate (S.true) of 0.8 and a tag-recovery rate of
0.1. You should inspect the matrix by clicking on the name in the Environment

window. The above parameters result in a vector of 1000, 800, and 640 tags at risk from a release in period 1. Cells in the matrix of fates are obtained as the number at risk multiplied by the tag-recovery rate, plus a final column for fish not seen again (Williams et al., 2002). The number released (1000) is arbitrary (and ambitious), but using a large value is helpful for reviewing the calculations that produce the two matrices. The parameter values can be changed to mimic any planned study design.

This section of code introduces a while loop. This different looping construct is needed because all rows except the last one include periods after the release. We need to be able to skip the additional processing in the last row, which has only the diagonal element. The final column for each row is the not-seen-again calculation. That value is highest for the most recent release, because most of those fish will still be alive. That uncertainty about whether fish are still at risk or have died is reduced by extending the number of periods for tag returns. Our example has the same number of releases and return periods, but see Appendix A.4 for an example with additional periods of tag returns.

Inspecting the matrix of fates is helpful for understanding how survival is estimated. The expected tag returns within any column form ratios that equal the survival rate. For example, we would expect 80 tags returned in the second period from the first release compared to 100 from the second release, because the first release has been at large for one additional period. Similarly the ratio from successive rows of returns in period 3 also equals the survival rate. Thus we have several cells (and ratios) that provide information about survival, which we assumed to be constant over the three periods.

The final lines of code in this section introduce the stochasticity associated with a multinomial distribution. The matrix of fates (expected values) provides the cell probabilities (once rescaled, internal to the rmultinom() function). It is instructive to compare the expected values (TagFate) and random realizations (RandRet) to see the level of variability associated with the multinomial distribution. Now the ratios of random values within a column no longer equal (but will vary around) the true survival rate.

```
# Load necessary library packages
library(rjags)
library(R2jags)

# Specify model in JAGS
sink("TagReturn.txt")
cat("
model {
  # Priors
  S.est ~ dunif(0,1) # Time-independent survival rate
  r.est ~ dunif(0,1) # Time-independent tag-recovery rate

  # Calculated value
```

```r
    A.est <- 1-S.est # Rate of mortality

# Cell probabilities
for (i in 1:Periods) {
    p.est[i,i] <- r.est
    for (j in (i+1):Periods) {
        p.est[i,j] <- p.est[i,(j-1)] * S.est
        } #j
    p.est[i,Periods+1] <- 1 - sum(p.est[i, i:Periods])
    # Last column is probability of not being seen again
    } #i

# Likelihood
  for (i in 1:Periods) {
    RandRet[i,i:(Periods+1)] ~ dmulti(p.est[i,i:(Periods+1)], NTags)
    }#i
  } # model

",fill=TRUE)
sink()

    # Bundle data
  jags.data <- list("RandRet", "Periods", "NTags")

  S.init <- runif(1,min=0,max=1)
  r.init <- 1-S.init  # Initialize r relative to S

  # Initial values
  jags.inits <- function(){list(S.est = S.init, r.est=r.init)}

  model.file <- 'TagReturn.txt'

  # Parameters monitored
  jags.params <- c("S.est", "r.est", "A.est")

   # Call JAGS from R
  jagsfit <- jags(data=jags.data, jags.params, inits=jags.inits,
                  n.chains = 3, n.thin=1, n.iter = 4000, n.burnin=2000,
                  model.file)
  print(jagsfit)
  plot(jagsfit)

  # Estimated posterior distributions
hist(jagsfit$BUGSoutput$sims.array[,,2], main="",
     xlab="Survival rate")
abline(v=S.true,  lty=3, lwd=3)
hist(jagsfit$BUGSoutput$sims.array[,,4], main="",
```

```
    xlab="Tag recovery rate")
abline(v=r.true,   lty=3,  lwd=3)
```

The JAGS analysis begins with the usual uninformative priors for the two rates. We calculate the total mortality rate (A.est) as the complement of the survival rate (1-S.est). One important advantage of a Bayesian analysis is that we automatically obtain measures of uncertainty for functions of model parameters. Here the function is simply 1-S.est, but this advantage applies regardless of the function's complexity.

The likelihood calculations (dmulti()) require cell probabilities for the multinomial distribution, so we calculate those probabilities directly rather than keeping track of numbers of tags by fate. The probability of a tag from release 'i' being returned in period 'j' is p.est[i,j] <- p.est[i,(j-1)] * S.est. Make sure that you understand this expression for specific values of 'i' and 'j'. Can you think of another way of coding this calculation? Note that for loops work differently in JAGS compared to R, so we are able to use them for row and column processing. In JAGS, the 'j' loop is not executed when i=3 because j>Periods.

JAGS run-time errors frequently occurred when independent uniform 0-1 distributions were used to obtain initial values for S.est and r.est. Using 1-S.init as the initial value for r.est appears to work well in practice. For this time-independent model, the tag recovery rate should be less than the total mortality rate due to the other possible outcomes (e.g., harvest not reported, natural death).

The last four lines of code plot the estimated posterior distributions for survival and tag-recovery rate, along with the true values (Figure 6.2). Estimates of survival and tag-recovery rates are typically reliable, which is not surprising given the very large releases and only two time-independent parameters to estimate. The estimated number of effective parameters was 1.9 for one run, which is consistent with the two apparent model parameters (S.est, r.est) and A.est, which is not a model parameter but simply a function of S.est.

An empirical example using male wood duck (*Aix sponsa*) band returns is included as Appendix A.4.

If you think about a local system with which you are familiar, how practical would it be to tag 1000 fish at the start of each period (e.g., annually)? If you were responsible for planning the tagging operation, what sampling gear would you recommend? Would there be any potential concern about mortality due to handling? What size field crew would be needed, and how much time would the tagging operation require? Could you use simulation to explore different sample sizes before recommending a specific field design?

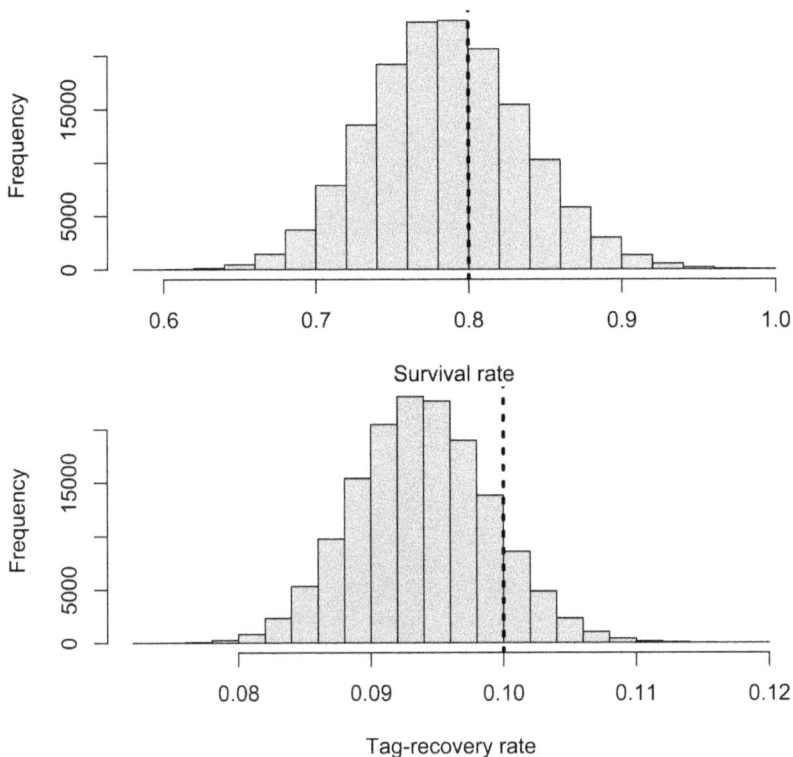

FIGURE 6.2 Example of estimated posterior distributions for survival and tag-recovery rates from 3-year tag-return study with annual releases of 1000 tagged fish. Dotted vertical lines indicate true values.

6.4 Exercises

1. For the default age-composition analysis without gear selectivity (Section 6.1), compare credible intervals for survival rate at sample sizes of 30 versus 200 fish.

2. For the telemetry method of estimating survival (Section 6.2), approximately what sample size would produce an estimate with the level of precision recommended by Robson and Regier (1964) for research (10%) studies?

3. For the tag-return method (Section 6.3), compare credible intervals for survival rate at releases of 50, 100, and 200 fish. How do the results compare to the Robson and Regier (1964) recommendations? In what ways is this simulated field study optimistic about the reliability of results?

7

Mortality Components

Reliable estimates of the survival rate (Chapter 6) are essential in understanding fish population dynamics, but for exploited species, there is generally interest in partitioning mortality (1-survival) into its component rates. Knowing the relative importance of fishing versus natural mortality clarifies the extent to which fishery managers can use regulations to affect population change. The sections that follow illustrate different approaches for examining the impact of fishing and for estimating fishing and natural mortality rates.

7.1 Exploitation rate

A tagging study can provide a real-time estimate of the exploitation rate, or the fraction of the population being harvested. Similar to the tag-return method of estimating survival (Section 6.3), the only field work required is the initial tagging effort. Here there is a single release of fish with external tags at the start of the study, and the number of returned tags is recorded over a specified interval. The study duration could be an entire year, or a fishing season for a population managed by seasonal harvest. A high exploitation rate might indicate that harvest regulations such as a size limit or creel limit could be worthwhile. Alternatively, a low estimate might indicate that other factors such as natural mortality or recruitment may be more important in regulating population size. With only a single release, we cannot estimate survival (or its complement, mortality), but it is nevertheless an important first step in judging the relative impact of fishing.

There are several practical issues to consider in planning an exploitation rate study. We want to ensure that essentially all tags of harvested fish are reported, so high-reward tags (e.g., $100) could be used (Pollock et al., 2001). Tag loss is another potential source of bias. It can be estimated by double-tagging (giving some or all fish two tags, and recording returns of two versus only one tag), but it greatly complicates the analysis. For this example, we will keep things simple and assume that tag loss is negligible. We also assume that there is no immediate tagging mortality (i.e., due to capture, handling and

tagging). This can be challenging to investigate. It is sometimes estimated by
holding a sample of fish in a cage after tagging but this can itself be a source of
mortality (Pollock and Pine, 2007). For now, we assume that there is no short-
or long-term mortality associated with tagging. We assume that all caught fish
are harvested. If catch-and-release was occurring, our model could be extended
to estimate rates of harvest and catch-and-release. The decision of how many
fish to tag will depend on the required precision of the study, and as usual
it can be examined through simulation. Finally, it is important that tagged
fish be representative of the entire population. This refers not only to size or
age, but also spatial distribution. Tagging could be done at randomly selected
locations, or when the entire study population is concentrated within a single
area (e.g., winter holding area or spawning location). The key is for all fish,
including the tagged subset, to have the same probability of being harvested.

Our simulation code uses an assumed exploitation rate (u) of 0.4. The value
could be based on a literature review or prior work from the study area or
similar areas regionally. The number of returned tags has a binomial distribution
(Section 3.1.2):

```
rm(list=ls()) # Clear Environment

# Tagging study to estimate exploitation rate
u.true <- 0.4
n.tagged <- 100
n.returned <- rbinom(n=1, prob=u.true, size=n.tagged)
```

The simulation provides only a single observation for the number of returned
tags. What lines of code could you add to look at the distribution of returned
tags (using a new variable for the vector of tag returns)?

The JAGS code is also extremely simple. We have one line of code for the
uninformative prior distribution for the exploitation rate, and one for the
likelihood.

```
# Load necessary library packages
library(rjags)
library(R2jags)

# JAGS code
sink("ExpRate.txt")
cat("
model{
    # Priors
```

```
      u.est ~ dunif(0,1)   # Exploitation rate

      # Likelihood
      n.returned ~ dbin(u.est, n.tagged)
}
      ",fill=TRUE)
sink()

# Bundle data
jags.data <- list("n.returned", "n.tagged")

# Initial values
jags.inits <- function(){ list(u.est=runif(n=1, min=0, max=1))}

model.file <- 'ExpRate.txt'

# Parameters monitored
jags.params <- c("u.est")

# Call JAGS from R
jagsfit <- jags(data=jags.data, jags.params, inits=jags.inits,
                n.chains = 3, n.thin=1, n.iter = 4000, n.burnin=2000,
                model.file)
print(jagsfit)
plot(jagsfit)
```

A single run with a release of 100 tagged fish provided a credible interval of
0.27-0.46 (from a random return of 36 tags). Experiment with different sample
sizes and consider what level of precision would be adequate for management
purposes. As always, this is an optimistic scenario, with the number of returned
tags perfectly conforming to a binomial distribution.

7.1.1 Informative prior distribution

Banner et al. (2020) discourage the (at least uncritical) use of "default" prior
distributions (which are used throughout this book). It is often true that
previous studies contain relevant information, and Bayesian methods provide
an objective framework for building on that previous information (McCarthy,
2007). The simple example of estimating the exploitation rate is a perfect
opportunity to look at how an informative prior would be implemented and to
explore its effect on the posterior distribution.

To compare posterior distributions obtained using uninformative and informative priors, we use two copies of the above code. In the uninformative prior version, we simply replace dunif(0,1) with dbeta(1,1) in the JAGS code, and we replace the uniform initial value with the corresponding beta (rbeta(n=1, shape1=1, shape2=1)). Neither change affects the calculations because the two distributions are equivalent, but it sets up the model for the informative prior version.

For the informative case, the exploitation rate simulation is unchanged. We simply add two lines of code for the pilot study that will be used for the prior. The informative beta distribution in the JAGS code has two parameters: successes+1 and failures+1 (Bolker, 2008). We augment jags.data with the pilot study data, and use initial values from the informative prior. The JAGS code is followed by R code for producing two plots, showing the prior and posterior distributions for the uninformative (upper plot) and informative (lower) cases. We specify margins for each plot to reduce blank space. The density() function provides a smoothed probability density for the exploitation rate estimates. The lines() function adds the prior distribution to each plot, generated using the dbeta() function.

```
rm(list=ls()) # Clear Environment

# Tagging study to estimate exploitation rate
u.true <- 0.4
n.tagged <- 100
n.returned <- rbinom(n=1, prob=u.true, size=n.tagged)

# Load necessary library packages
library(rjags)
library(R2jags)

# JAGS code
sink("ExpRate.txt")
cat("
model{
    # Priors
    u.est ~ dbeta(1,1)   # Exploitation rate

    # Likelihood
    n.returned ~ dbin(u.est, n.tagged)
}
    ",fill=TRUE)
sink()
```

```r
# Bundle data
jags.data <- list("n.returned", "n.tagged")

# Initial values
jags.inits <- function(){ list(u.est=rbeta(n=1, shape1=1, shape2=1))}

model.file <- 'ExpRate.txt'

# Parameters monitored
jags.params <- c("u.est")

# Call JAGS from R
jagsfit <- jags(data=jags.data, jags.params, inits=jags.inits,
                n.chains = 3, n.thin=1, n.iter = 4000, n.burnin=2000,
                model.file)
print(jagsfit)
plot(jagsfit)

# Informative prior
n.pilot <- 20
n.returned.pilot <- rbinom(n=1, prob=u.true, size=n.pilot)

# Load necessary library packages
library(rjags)
library(R2jags)

# JAGS code
sink("ExpRate.txt")
cat("
model{
    # Priors
    u.est ~ dbeta((n.returned.pilot+1),
        (n.pilot-n.returned.pilot+1)) # Exploitation rate

    # Likelihood
    n.returned ~ dbin(u.est, n.tagged)
}
    ",fill=TRUE)
sink()

# Bundle data
jags.data <- list("n.returned", "n.tagged", "n.pilot",
                  "n.returned.pilot")
```

```
# Initial values
jags.inits <- function(){ list(u.est=rbeta(n=1,
                        shape1=(n.returned.pilot+1),
                        shape2=(n.pilot-n.returned.pilot+1)))}


model.file <- 'ExpRate.txt'

# Parameters monitored
jags.params <- c("u.est")

# Call JAGS from R
informative <- jags(data=jags.data, jags.params, inits=jags.inits,
             n.chains = 3, n.thin=1, n.iter = 4000, n.burnin=2000,
             model.file)
print(informative)
plot(informative)

# Vague prior and posterior
par(mfrow=c(2,1))
par(mar=c(2,4,2,1)) # bottom, left, right, top margins
dens1 <- density(jagsfit$BUGSoutput$sims.list$u.est)
dens2 <- density(informative$BUGSoutput$sims.list$u.est)
y.bounds <- range(dens1$y, dens2$y)
x.bounds <- range(jagsfit$BUGSoutput$sims.list$u.est,
             informative$BUGSoutput$sims.list$u.est)
plot(density(jagsfit$BUGSoutput$sims.list$u.est), xlab="", main="",
     xlim=x.bounds, ylim=y.bounds)
abline(v=u.true,  lty=3, lwd=3)
x.vec <- seq(x.bounds[1], x.bounds[2], length.out=101)
lines(x.vec,
      dbeta(seq(x.bounds[1], x.bounds[2], length.out=101), shape1=1,
            shape2=1), lty=2)
legend(x = "topleft", lty = c(2,1), text.font = 1,
       legend=c("Prior", "Posterior"))

# Informative prior and posterior
par(mar=c(4,4,1,1))
plot(dens2,xlab="Estimated exploitation rate", main="",
     xlim=x.bounds, ylim=y.bounds)
abline(v=u.true,  lty=3, lwd=3)
lines(x.vec, dbeta(seq(x.bounds[1], x.bounds[2],
                   length.out=101), shape1=(n.returned.pilot+1),
                   shape2=(n.pilot-n.returned.pilot+1)), lty=2)
```

A pilot study provides a valuable opportunity to work out field logistics, and allows for construction of an informative prior. Informative priors can also be based on the mean and variance of published estimates, and they can include the additional variation within individual studies (McCarthy and Masters, 2005; Regehr et al., 2018). Another approach is to base the prior on ecological theory, for example, a prior for survival rate based on its relationship to body mass [(McCarthy and Masters, 2005). For our tagging example, the beta distribution worked well as an informative prior because the two parameters are based on the number of successes (returned tag) and failures (not returned). The plotted distributions suggest a modest gain in precision from a pilot release of 20 tagged fish, given a full study release of 100 (Figure 7.1). The effect of an informative prior can also be evaluated by comparing the width of credible intervals (McCarthy and Masters, 2005).

The benefit of an informative prior depends on the information content of the prior versus the new data (McCarthy and Masters, 2005). Lemoine (2019) recommends carrying out analyses of real data using multiple prior distributions of varying strengths, to better understand the information content of the new data and the effect of priors. One other point to consider if conducting a pilot-tagging study, rather than using published information, is the random aspect of the number returned. Try multiple runs with small pilot releases to see how often the prior is unhelpful for estimating the true exploitation rate.

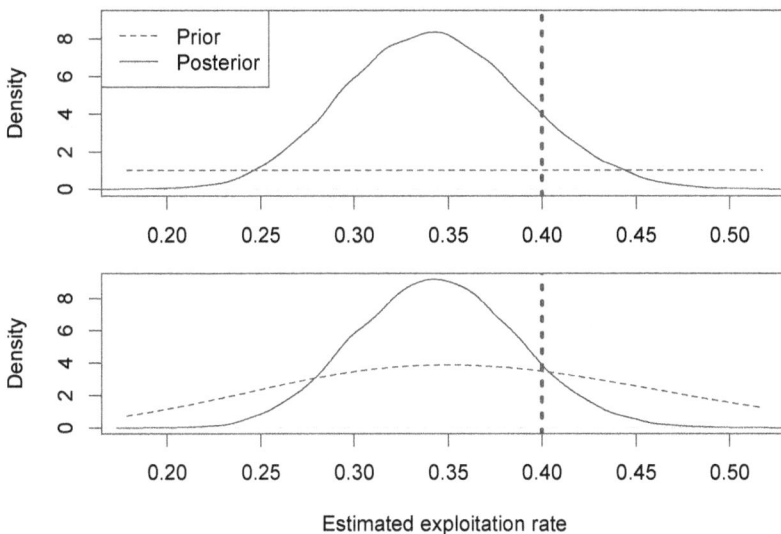

FIGURE 7.1 Prior and posterior distributions for estimated exploitation rate, using uninformative (upper) and informative (lower) priors. The full study sample size is 100 tagged fish; the pilot study (lower panel) has a sample size of 20. The true exploitation rate is shown as a dotted vertical line.

What factors would you consider when deciding whether to base an informative prior on a pilot study versus a summary of published estimates?

7.2   Completed tagging study

A single release of tags (Section 7.1) can also be used to estimate the rate of natural mortality. Hearn et al. (1987) provide a method that can be used in completed tagging studies, that is, a study of sufficient duration that tag returns have ceased (assumed to indicate that no live tagged fish remain). Their model uses instantaneous rates, which we have up to this point avoided. The relationship between probabilities (survival, death, exploitation, natural death) and instantaneous rates is as follows. The instantaneous rate of population decline is $dN/dt = -Zt$, where N is population size and Z is the instantaneous total mortality rate (rate of decline due to fishing and natural sources). An advantage of instantaneous rates is that they are additive. Stock assessment models routinely partition Z into whatever categories are useful and estimable; for example, fishing versus natural mortality, commmercial versus recreational fishing, longline versus trawl, etc. The most common partitioning is for fishing (F) versus natural (M) sources: $dN/dt = -(M + F)t$. Integrating this equation provides us with an expression for population size at any time t:

$$N_t = N_0 * exp(-(M + F) * t)$$

where N_0 is initial abundance, and $S = exp(-(M + F) * t)$ is the survival rate. The total mortality rate (probability of dying from all causes) is $A = 1 - S$. The exploitation rate is obtained as u=FA/Z, which makes sense as the fraction of total mortality due to fishing (F/Z). Similarly the probability of natural death is v=MA/Z. These expressions make it possible to go back and forth between probabilities and instantaneous rates, depending on which is more useful or traditional for a particular model. These expressions apply for the most common situation where fishing and natural mortality occur simultaneously (referred to by Ricker (1975) as a Type 2 fishery). Ricker (1975) provides alternate expressions for the less common case (a Type 1 fishery) where natural mortality occurs after a short fishing season.

Unlike the probabilities for survival, harvest, etc., instantaneous rates are unbounded. In a Bayesian context, this makes them more challenging to estimate compared to probabilities, which are conveniently bounded by 0 and 1. Instantaneous rates are also less intuitive in terms of their magnitude compared to probabilities. There is typically more uncertainty about M than F because natural deaths are almost never observed (Hearn et al., 1987; Quinn and Deriso, 1999). The joke about M is that it is often assumed to be 0.2, because .2 looks like a distorted version of a question mark. In a review of published natural mortality estimates, Vetter (1988) found that a majority were less than 0.5, but values exceeding 2.0 were occasionally reported.

The Hearn et al. (1987) method requires only that fishing mortality in each period be greater than zero, in order to establish the study's endpoint. Their model assumes a constant rate of natural mortality, so the estimated rate would in practice represent the average over the study. Their base model assumes that tags of harvested fish are always reported, there is no mortality associated with tagging, and there is no tag loss.

The Hearn et al. (1987) model used exact times at large for returned tags to estimate the natural mortality rate. The following Bayesian version is a modification of their method, using tag returns by period (assumed here to be annual) to establish the length of the completed study. Simulated fishing mortality rates are greater than 0 but vary randomly within a user-defined range. We begin with simulation code:

```
rm(list=ls()) # Clear Environment

n.tagged <- 100
Periods <- 40
# Set to high value to ensure that end of study is observed
F.true <- runif(n=(Periods-1), min=0.3, max=0.5)
M.true <- 0.2
S.true <- exp(-F.true - M.true)
u.true <- F.true * (1-S.true) / (F.true + M.true)

TagsAtRisk <- array(data=NA, dim=Periods)
TagsAtRisk[1] <- n.tagged
TagFates <- array(data=NA, dim=(Periods-1)) # Returns by period

for (j in 2:Periods){
   TagsAtRisk[j] <- rbinom(n=1, size=TagsAtRisk[j-1], prob=S.true[j-1])
   TagFates[j-1] <- rbinom(n=1, size=(TagsAtRisk[j-1]
                      -TagsAtRisk[j]),prob=(F.true[j-1]
                      /(F.true[j-1]+M.true)))
   # Random realization: fishing deaths
} #j

t.Completed <- max(which(TagFates[]>0))
# Locate period when study completed (last return)
NatDeaths <- n.tagged - sum(TagFates)
TagFates <- c(TagFates[1:t.Completed],NatDeaths)
# Returns + all natural deaths
```

The vectors for tags-at-risk and fates are binomially-distributed random realizations. Tags at risk are obtained using the number at risk in the previous period and the survival rate. Tag returns (stored in tag fates vector) are obtained from the number of deaths between periods and the fraction of deaths due to fishing (F/Z). The which() function provides a vector of periods when tag returns are greater than 0, and the max() function chooses the final period with at least one tag return. Try which(TagFates[]>0) in the Console to make clear how this expression works. The final version of the tag fates vector contains tag returns for the completed study length plus a final value for natural deaths (any tagged fish not seen in the completed tagging study).

The JAGS code for fitting the model is similar to that for the tag-based method of estimating survival (Section 6.3), except that our parameters now are instantaneous rates. We use an arbitrary uniform prior distribution, with an upper bound (2) set high enough to ensure that the interval contains the true value. The likelihood uses a multinomial distribution (Section 3.1.3), where cell probabilities represent the fraction returned in each period plus a final value for the fraction not seen again (natural deaths in our completed tagging study). The last three lines of code show the estimated posterior distribution compared to the true value.

```
# Load necessary library packages
library(rjags)
library(R2jags)

# JAGS code
sink("Hearn.txt")
cat("
  model {
  # Priors
  M.est ~ dunif(0, 2)   # Instantaneous natural mortality rate
  for (i in 1:t.Completed) {
     F.est[i] ~ dunif(0,2) # Instantaneous fishing mortality rate
     S.est[i] <- exp(-M.est - F.est[i])
     u.est[i] <- F.est[i] * (1-S.est[i])/(F.est[i]+M.est) # FA/Z
  } #i

# Cell probabilities
  p.est[1] <- u.est[1]
  for (i in 2:t.Completed) {
    p.est[i] <- prod(S.est[1:(i-1)])*u.est[i]
     } #i
    p.est[t.Completed+1] <- 1 - sum(p.est[1:t.Completed])
      # Prob of not being seen again
```

```
  TagFates[1:(t.Completed+1)] ~ dmulti(p.est[1:(t.Completed+1)],
                            n.tagged)

  }
  ",fill=TRUE)
sink()

# Bundle data
jags.data <- list("TagFates", "t.Completed", "n.tagged")

# Initial values
jags.inits <- function(){list(M.est=runif(1, 0, 2),
                              F.est=runif(t.Completed, 0, 2))}


model.file <- 'Hearn.txt'

# Parameters monitored
jags.params <- c("M.est", "F.est")

# Call JAGS from R
jagsfit <- jags(data=jags.data, jags.params, inits=jags.inits,
                n.chains = 3, n.iter = 4000, n.burnin=2000,
                model.file)
print(jagsfit)
plot(jagsfit)


par(mfrow=c(1,1)) # Default plot settings
mar = c(5.1, 4.1, 4.1, 2.1)
# Look at posterior distribution versus true value
hist(jagsfit$BUGSoutput$sims.list$M.est,
     xlab="Estimated natural mortality rate", main="")
abline(v=M.true,  lty=3, lwd=3)
```

This approach works because there are only two fates for tagged fish: caught and reported or natural death. Using a completed tagging study means that (when all assumptions hold) the number of natural deaths is known exactly (number of tag releases - total returns). The model adjusts the estimate of M in order to account for all the natural deaths by the end of the completed study. The length of the completed tagging study depends on the assumed values for F and M. The estimated natural mortality rate will be higher when the study length is short and when many tagged fish are not seen again.

There are elements of randomness in the length of the completed tagging study. One example is when tagged fish at risk survive for one or more periods beyond the last tag return but are ultimately natural deaths. This can occasionally be

seen by comparing the TagsAtRisk and TagFates vectors. There can also be
embedded zeros in the vector of tag returns when the number of tags at risk is
low. Simulation runs suggest that this approach to determining study length
works reasonably well but should be evaluated for the likely range of fishing
and natural mortality rates.

For the default settings, estimates of M tend to be reliable (Figure 7.2), which
is unsurprising given that it is modeled as a constant rate and estimated from
multiple years of data. Run the full code (simulation and analysis) multiple
times, using different assumed values for the natural mortality rate, to build
intuition about the reliability of this method. (Periods here are assumed to be
years, but a larger number of periods might be needed for monthly intervals
or if a very low natural mortality rate is used.) How would uncertainty change
with a different number of tags released?

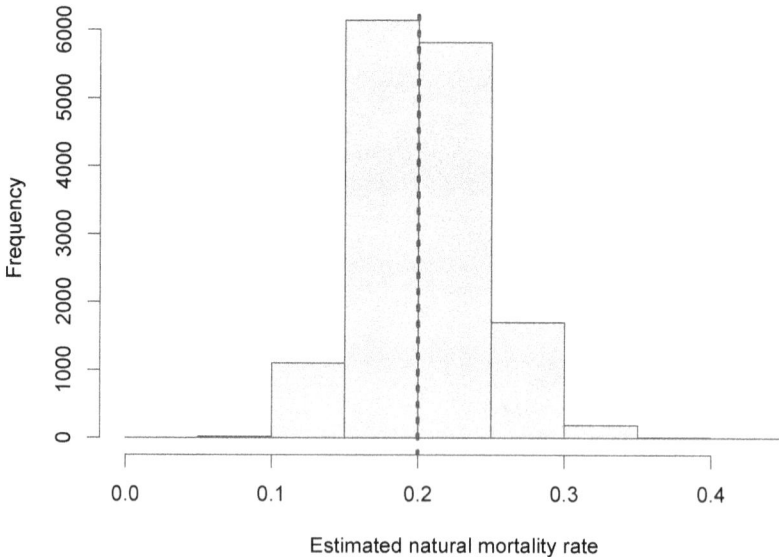

FIGURE 7.2 Example of estimated posterior distribution and true natural
mortality rate (dotted vertical line) using the completed tagging study (Hearn
et al. 1987) approach.

Estimated fishing mortality rates for the first few periods are moderately
accurate, especially for larger tag releases. Estimates near the end of the
completed study are typically high with wide credible intervals. There would
likely be some improvement from simplifying the model to estimate only an
average annual fishing mortality rate rather than period-specific estimates.

7.3 Multi-year tagging study

We return to the multi-year tagging study (Section 6.3) but now assume that we have information about the tag-reporting rate. This information allows us to partition mortality into fishing and natural sources. Tag-reporting rate is fixed at 1.0 in the following code, as might be appropriate when using high-reward tags. In a later section of code, we estimate it internally, using planted tags (Hearn et al., 2003) and an additional likelihood component. Information about the tag-reporting rate is very powerful. For example, a 100% tag-reporting rate means that (assuming no assumptions are violated) you ultimately know the number of natural deaths because all other tags will be reported (as in Section 7.2). There is uncertainty in the short term, because tags not returned could belong to fish still alive and at risk; so information about natural deaths is clear only after sufficient time has passed.

Our simulation code is similar to that of Section 6.3 except that we now specify the probabilities of harvest (u.true) and natural death (v.true). We use vectors so that the rates can vary by period. We have arbitrarily set the probabilities for harvest higher than for natural death, to investigate whether we can reliably detect the predominant threat in our simulated study design.

```
rm(list=ls()) # Clear Environment

# Parameter values for three-yr experiment
Periods <- 3  # Equal number of release and return periods assumed
NTags <- 1000 # Number released each period
NumTagged <- rep(NTags, Periods) # Release NTags fish each period

u.true <- c(0.4, 0.25, 0.3) # Probability of fishing death
v.true <- c(0.1, 0.15, 0.2) # Probability of natural death
lam.true <- 1  # Reporting rate

# Calculated value(s), array creation
S.true <- 1 - u.true - v.true
TagsAtRisk <- array(data=NA, dim=c(Periods, Periods))
TagFate <- array(data=NA, dim=c(Periods, Periods+1))
   # Extra column for tags not seen again

for (i in 1:Periods){
# Expected values for tags-at-risk, returned/not-seen-again
   TagsAtRisk[i,i] <- NumTagged[i]
   TagFate[i,i] <-  TagsAtRisk[i,i] * lam.true * u.true[i]
    # Returns in release period
```

```
  j <- i+1
  while(j <= Periods) {
    TagsAtRisk[i,j] <- TagsAtRisk[i,(j-1)]*S.true[j-1]
    TagFate[i,j] <- TagsAtRisk[i,j] * lam.true * u.true[j]
    j <- j+1
  } #while
  TagFate[i, Periods+1] <- NTags-sum(TagFate[i,i:Periods])
    # Tags not seen again
  } #i

RandRet <- array(data=NA, dim=c(Periods, Periods+1))
  # Extra column for fish not seen again
#Random trial
  for (i in 1:Periods) # Upper diagonal matrix of random tag returns
  {
    RandRet[i,i:(Periods+1)] <- t(rmultinom(1, NumTagged[i],
                                  TagFate[i,i:(Periods+1)]))
  } #i
```

The analysis is again similar to Section 6.3 except that now we have a multi-nomial distribution (Section 3.1.3) with three possible fates (survival, harvest, natural death). Following Section 9.6 of Kéry and Schaub (2012), we use three gamma-distributed (Section 3.2.5) parameters that are scaled by their sum, thus ensuring that the rescaled values sum to 1. The parameters of interest (u, v, S) are calculated functions of the 'a' parameters, which are used internally but not themselves of interest.

```
# Load necessary library packages
library(rjags)
library(R2jags)

  # JAGS code
  sink("TagReturn.txt")
  cat("
model {
# Priors
lam.est <- 1 #  100% reporting of tags
for (i in 1:Periods) {
    for (j in 1:3) # Rows are return periods, columns are fates
      {
        a[i,j] ~ dgamma(1,1)
      } #j
    S.est[i] <- a[i,1]/sum(a[i,]) # Probability of survival
    u.est[i] <- a[i,2]/sum(a[i,]) # Probability of harvest
```

```
      v.est[i] <- a[i,3]/sum(a[i,]) # Probability of natural death
      # v.est[i] could also be obtained as 1-S.est[i]-u.est[i]
   } #i

# Cell probabilities
   for (i in 1:Periods) {
     p.est[i,i] <- lam.est * u.est[i]
     for (j in (i+1):Periods) {
       p.est[i,j] <- prod(S.est[i:(j-1)])*lam.est*u.est[j]
       } #j
     p.est[i,Periods+1] <- 1 - sum(p.est[i, i:Periods])
     # Probability of not being seen again
     } #i
   for (i in 1:Periods) {
   RandRet[i,i:(Periods+1)] ~ dmulti(p.est[i,i:(Periods+1)],
                                    NumTagged[i])

   }#i
   }
   ",fill=TRUE)
   sink()


# Bundle data
   jags.data <- list("RandRet", "Periods", "NumTagged")

# Initial values
   a.init <- array(data=NA, dim=c(Periods, 3))
   for (i in 1:3){
     a.init[i,] <- runif(n=3, min=0, max=1)
     a.init[i,] <- a.init[i,]/sum(a.init[i,]) # Rescale to sum to 1
   }
   jags.inits <- function(){ list(a=a.init)}

   model.file <- 'TagReturn.txt'

   # Parameters monitored
   jags.params <- c("u.est", "v.est")

    # Call JAGS from R
   jagsfit <- jags(data=jags.data, jags.params, inits=jags.inits,
                   n.chains = 3, n.thin = 1, n.iter = 4000,
                   n.burnin=2000, model.file)
   print(jagsfit)
   plot(jagsfit)
```

We provide arbitrary starting values for the 'a' parameters by rescaling uniform (0,1) random variates. Using a large release on each occasion (1000), our estimates are generally close to the true values except that v.est[3] is imprecise and typically overestimated. It is the most difficult time-dependent parameter to estimate. We get direct information (from returned tags) about the final exploitation rate, but there is much uncertainty about whether fish not seen are still at risk or are natural deaths. This is especially true for the final release because most fish are still at large (examine the TagFate and RandRet matrices). Note that this differs from a completed tagging study (Section 7.2), where the study continues until no live fish remain.

The analysis can be extended to include estimation of the tag-reporting rate. We use the simplest approach of a binomial experiment using planted tags (Hearn et al., 2003). The fraction of planted tags that is returned is a direct estimate of the tag-reporting rate. The only change required in the simulation code is to replace the assignment lam.true <- 1.0 with the following three lines:

```
lam.true <- 0.9   # Reporting rate
PlantedTags <- 30
PlantReturns <- rbinom(n=1, size=PlantedTags, prob=lam.true)
```

The assumed tag-reporting rate and number of planted tags are arbitrary and can be varied to see the effect on parameter uncertainty.

```
# Load necessary library packages
library(rjags)
library(R2jags)

# JAGS code
  sink("TagReturn.txt")
  cat("
model {
# Priors
lam.est ~ dunif(0,1) # Reporting rate
for (i in 1:Periods) {
    for (j in 1:3) # Rows are return periods, columns are fates
      {
        a[i,j] ~ dgamma(1,1)
      } #j
    S.est[i] <- a[i,1]/sum(a[i,]) # Probability of survival
    u.est[i] <- a[i,2]/sum(a[i,]) # Probability of harvest
    v.est[i] <- a[i,3]/sum(a[i,]) # Probability of natural death
    # v.est[i] could also be obtained as 1-S.est[i]-u.est[i]
  } #i
```

```
# Cell probabilities
  for (i in 1:Periods) {
    p.est[i,i] <- lam.est * u.est[i]
    for (j in (i+1):Periods) {
      p.est[i,j] <- prod(S.est[i:(j-1)])*lam.est*u.est[j]
      } #j
    p.est[i,Periods+1] <- 1 - sum(p.est[i, i:Periods])
      # Prob of not being seen again
    } #i
  for (i in 1:Periods) {
  RandRet[i,i:(Periods+1)] ~ dmulti(p.est[i,i:(Periods+1)],
                              NumTagged[i])
  }#i
  PlantReturns ~ dbin(lam.est, PlantedTags)
    # Additional likelihood component
}
  ",fill=TRUE)
  sink()

# Bundle data
  jags.data <- list("RandRet", "Periods", "NumTagged", "PlantedTags",
                  "PlantReturns")

# Initial values
  a.init <- array(data=NA, dim=c(Periods, 3))
  for (i in 1:3){
    a.init[i,] <- runif(n=3, min=0, max=1)
    a.init[i,] <- a.init[i,]/sum(a.init[i,])  # Rescale to sum to 1
  }
  jags.inits <- function(){ list(a=a.init)}

  model.file <- 'TagReturn.txt'

  # Parameters monitored
  jags.params <- c("lam.est", "u.est", "v.est")

  # Call JAGS from R
  jagsfit <- jags(data=jags.data, jags.params, inits=jags.inits,
                  n.chains = 3, n.thin = 1, n.iter = 10000,
                  n.burnin=5000, model.file)
  print(jagsfit)
  plot(jagsfit)
```

The auxiliary planted tag experiment requires only a few changes in the JAGS code. We provide an uninformative prior distribution for the now estimated parameter lam.est, and pass in the number of planted tags and returns. We let JAGS generate an initial value for lam.est. The additional likelihood component describes the binomial experiment. Convergence may be slower for this version of the code, so the number of MCMC iterations was increased to 10,000.

Estimates have similar precision to the original example if the reporting rate is high (Figure 7.3). Precision is reduced at lower reporting rates because of the reduced sample size of returned tags and uncertainty about lambda.

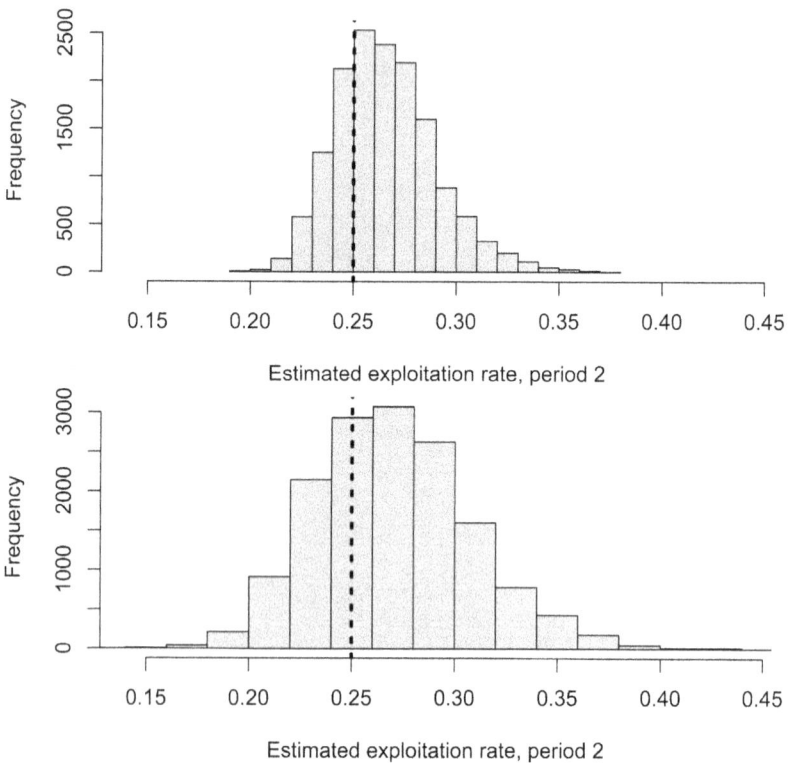

FIGURE 7.3 Estimated posterior distributions for exploitation rate in period 2 at tag reporting rates of 0.9 (top) and 0.5 (bottom), using a multi-year tagging study with planted tags. The true exploitation rate for period 2 is shown as a dotted vertical line.

7.4 Telemetry-based

In Section 6.2, we used fixed receiving stations and telemetry tags to estimate the survival rate. Here we extend that approach to partition total mortality into fishing and natural components by giving telemetered fish an external high-reward tag (Design C of Hightower and Harris, 2017). Live fish "detect themselves" and provide information on survival by passing within range of a receiver. Harvest information comes from the reported high-reward tags. Thus we have observations on two of the three possible states, with natural mortality being the only unobserved true state. An important practical advantage of this approach is that field surveys are not needed after tagging is done (other than maintaining the receiver array and downloading detection data).

We begin with a detour to introduce dcat, the JAGS categorical distribution, used to estimate probabilities for observations from different categories.

```
rm(list=ls()) # Clear Environment

# Load necessary library packages)
library(rjags)
library(R2jags)

# JAGS code
sink("dcat_example.txt")
cat("
model {

# Priors
    for (j in 1:3) {
        a[j] ~ dgamma(1,1)
        } #j
    p[1] <- a[1]/sum(a[])
    p[2] <- a[2]/sum(a[])
    p[3] <- a[3]/sum(a[]) # Could also be obtained as 1-p[1]-p[2]

# Likelihood
  for (i in 1:N){
    y[i] ~ dcat(p[])
  } #i
}
    ",fill = TRUE)
sink()

# Bundle data
```

```r
y <- c(1, 1, 3, 3, 2) # Each observation drawn from category 1, 2, or 3
N <- length(y)
jags.data <- list("N", "y")

model.file <- 'dcat_example.txt'

# Parameters monitored
jags.params <- c("p")

# Call JAGS from R
jagsfit <- jags(data=jags.data, jags.params, inits=NULL,
                n.chains = 3, n.thin=1, n.iter = 2000,
                model.file)
print(jagsfit)
plot(jagsfit)
```

Our dcat example has a 'y' vector with five observations in the range of 1 to 3. The categories can be of any sort; for example, age or species code for individual fish. The model parameters of interest are calculated variables representing the probabilities of the three categories. The probabilities sum to 1, so, as in Section 7.3, we calculate them using an 'a' vector with an uninformative gamma prior distribution. (A Dirichlet distribution can also be used as an uninformative prior distribution for probabilities that must sum to 1 (McCarthy, 2007).) For the purposes of this section, the 'p' vector could represent probabilities of survival, harvest, or natural death. The MCMC process adjusts the 'p' estimates to achieve the best possible match between model predictions and the observations ('y' vector). For simplicity, we let JAGS provide the initial values for the 'a' vector. The estimates have wide credible intervals because of the small sample size.

Returning to the telemetry example, we begin with simulation code, using arbitrary values for the number of occasions and sample size. Varying the patterns for rates of exploitation and natural death is helpful in judging the model's ability to detect moderate time variation. This is a multi-state model (Kéry and Schaub, 2012), with a matrix (PSI.STATE) describing the various true states and the probabilities for transitioning between states over time. The three rows represent true states alive, harvest, and natural death. A fish that is alive (row 1) can remain alive (probability S.true), be harvested (probability u.true), or become a natural mortality (probability v.true). A harvested fish (row 2) remains in that state with probability 1, as does a natural death (row 3). Note that the third matrix dimension for PSI.STATE is time, as the rates are time-dependent.

The observation matrix (PSI.OBS) has a similar structure. Rows are true states, columns are observed states, and the third dimension is time. There are three possible observed states: detected alive, harvested, and not detected. A live fish (row 1) can either be detected or not, with probabilities p[t] and 1-p[t]. A harvested fish (row 2) is assumed to be always detected (100% tag-reporting rate), and a natural death (row 3) is never detected.

The function for generating the simulated capture-history matrix (simul.ms) is complex, but key features are as follows. All fish are assumed to be tagged (transmitters and external tags) at time 1. Next we loop over occasions and use a multinomial trial to determine the true state transition between time t-1 and t (i.e., a live fish at time t-1 can remain alive, be harvested, or transition to natural death). To better understand the multinomial trial, try a simpler example in the Console. For example, rmultinom(n=1, size=1, prob=c(0.7, 0.25, 0.05)) is a single trial (n=1) for one individual (size=1), with three states at the specified probabilities. The function returns a matrix with a 1 in a randomly determined row. The which() function locates the '1' and returns the row index (= fish's state). This example could represent probabilities that a live fish at time t-1 survives (0.7), is harvested (0.25), or is a natural death (0.05). Returning to the code, a second multinomial trial based on the fish's true state draws a random observed outcome (e.g., a live fish is either detected or missed).

```
# Code modified from Kéry and Schaub 2012, section 9.5

rm(list=ls()) # Clear Environment

# Generation of simulated data
# Define probabilities as well as number of occasions,
# states, observations and released individuals
n.occasions <- 5
n.tagged <- 100   # Simplify from Kéry's version,
                  # only an initial release, and no spatial states

u.true <- c(0.2, 0.1, 0.3, 0.2)
v.true <- c(0.1, 0.15, 0.15, 0.1)
S.true <- 1-u.true-v.true
p <- c(1, 0.8, 0.8, 0.6, 0.7)
#  Detection probability fixed at 1 for first period,
# n.occasions-1 searches (e.g start of period 2, 3, ..)

n.states <- 3
n.obs <- 3

# Define matrices with survival, transition and detection probabilities

# 1. State process matrix, dimensions 1=state[t], 2=state[t+1], 3=time
PSI.STATE <- array(NA, dim=c(n.states, n.states, n.occasions-1))
    for (t in 1:(n.occasions-1)){
        PSI.STATE[,,t] <- matrix(c(
```

```
        S.true[t], u.true[t], v.true[t],
        0, 1, 0,
        0, 0, 1), nrow = n.states, byrow = TRUE)
        } #t

# 2.Observation process matrix.  1=true state, 2=observed state, 3=time
PSI.OBS <- array(NA, dim=c(n.states, n.obs, n.occasions))
    for (t in 1:n.occasions){
        PSI.OBS[,,t] <- matrix(c(
        p[t], 0, 1-p[t],
        0,  1, 0,  # Caught w/ high-reward tag, assume 100% reporting
        0, 0, 1), nrow = n.states, byrow = TRUE)
          # Natural deaths never detected
        } #t

# Define function to simulate multistate capture-recapture data
simul.ms <- function(PSI.STATE, PSI.OBS, n.tagged, n.occasions){
    CH <- CH.TRUE <- matrix(NA, ncol = n.occasions, nrow = n.tagged)
    CH[,1] <- CH.TRUE[,1] <- 1 # All releases at t=1
    for (i in 1:n.tagged){
      for (t in 2:n.occasions){
          # Multinomial trials for state transitions
          CH.TRUE[i,t] <- which(rmultinom(n=1, size=1,
                          prob=PSI.STATE[CH.TRUE[i,t-1],,t-1])==1)
          # which vector element=1; i.e., which state gets random draw
          # at time t given true state at t-1. True state determines
          # which row of PSI.STATE provides probabilities

          # Multinomial trials for observation process
          CH[i,t] <- which(rmultinom(1, 1, PSI.OBS[CH.TRUE[i,t],,t])==1)
          # which observation gets the 1 random draw, given
          # true time t state.
          } #t
      } #i

   return(list(CH=CH, CH.TRUE=CH.TRUE))
      # True (CH.TRUE) and observed (CH)
   } # simul.ms
```

The JAGS code is also complex but does generally mirror the simulation code.

```
# Load necessary library packages
library(rjags)
library(R2jags)

# JAGS code for estimating model parameters
```

```
sink("Mort_Tel_HRTag.txt")
cat("
model {

# Priors
for (t in 1:(n.occasions-1)) {
    for (j in 1:3) { # Rows are return periods, columns are fates
        a[t,j] ~ dgamma(1,1)
        } #j
    S.est[t] <- a[t,1]/sum(a[t,]) # Probability of survival
    u.est[t] <- a[t,2]/sum(a[t,]) # Probability of harvest
    v.est[t] <- a[t,3]/sum(a[t,]) # Probability of natural death
    # v.est[t] could also be obtained as 1-S.est[t]-u.est[t]
    } #t

  p[1] ~ dbern(1)
  for (t in 2:n.occasions){
    p[t] ~ dunif(0,1)
    }#t

# Define state-transition and observation matrices
# Probabilities of true state at t+1 given state at time t
    for (t in 1:(n.occasions-1)){
    ps[1,t,1] <- S.est[t]
    ps[1,t,2] <- u.est[t]
    ps[1,t,3] <- v.est[t]
    ps[2,t,1] <- 0
    ps[2,t,2] <- 1
    ps[2,t,3] <- 0
    ps[3,t,1] <- 0
    ps[3,t,2] <- 0
    ps[3,t,3] <- 1
    } #t

# Probabilities of observed states given true state
    for (t in 2:n.occasions){
    po[1,t,1] <- p[t]   # Row 1 true state = alive
    po[1,t,2] <- 0
    po[1,t,3] <- 1-p[t]
    po[2,t,1] <- 0      # Row 2 true state = harvested
    po[2,t,2] <- 1
    po[2,t,3] <- 0
    po[3,t,1] <- 0      # Row 3 true state= natural death
    po[3,t,2] <- 0
    po[3,t,3] <- 1
    } #t

    # Likelihood
```

```
    for (i in 1:n.tagged){
    z[i,1] <- y[i,1] # Latent state known at time of tagging (t=1)
    for (t in 2:n.occasions){
    # State process: draw state at time t given state at time t-1
    z[i,t] ~ dcat(ps[z[i,(t-1)], (t-1),])
    # Observation process: obs state given true state at time t
    y[i,t] ~ dcat(po[z[i,t], t,])
    } #t
    } #i
}
    ",fill = TRUE)
sink()

# Generate initial values for z. Modified from Kéry code for
# JAGS inits, age-specific example 9.5.3
z.init <- function(ch) {
  # State 1=Obs 1 (alive) State 2=Obs 2 (fishing death).
  # Start w/ known states from obs capture-history,
  # replace "3" with possible state(s)
  ch[ch==3] <- -1  # Not observed so temporarily replace w/ neg value
  ch[,1] <- NA  # Initial value not needed for release period
  for (i in 1:nrow(ch)){
    if(max(ch[i,], na.rm=TRUE)<2){
      ch[i, 2:ncol(ch)] <- 1
      # Not detected dead so initialize as alive (after release period)
    } else {
    m <- min(which(ch[i,]==2))
      # Period when fishing death first detected
    if(m>2) ch[i, 2:(m-1)] <- 1
    # Initialize as alive up to period prior to harvest
    ch[i, m:ncol(ch)] <- 2
      # Initialize as dead after detected fishing death
    } # if/else
} # i
  return(ch)
} # z.init

# Call function to get simulated true (CH.TRUE) and obs (CH) states
sim <- simul.ms(PSI.STATE, PSI.OBS, n.tagged, n.occasions)
y <- sim$CH

# Bundle data
jags.data <- list("n.occasions", "n.tagged", "y")

# Initial values.
jags.inits <- function(){ list(z = z.init(y))}


model.file <- 'Mort_Tel_HRTag.txt'
```

```
# Parameters monitored
jags.params <- c("u.est", "v.est", "p")

# Call JAGS from R
jagsfit <- jags(data=jags.data, jags.params, inits=jags.inits,
                n.chains = 3, n.thin=1, n.iter = 5000,
                model.file)
print(jagsfit)
plot(jagsfit)
```

The parameters for rates of survival, exploitation and natural death use the same approach as in the above dcat example. The 'a' matrix is estimated internally using a vague gamma prior distribution, then the three fractions estimate the probabilities of the three possible true states (survival, harvest, natural death), which sum to 1. The latent true states ('z' matrix) are estimated in the likelihood section of the code. The dcat (categorical) distribution is used to draw the true state at time t given state at time t-1, as well as the observation at each time given the inferred true state.

One difficult part of fitting this model is obtaining initial values for the latent true states. JAGS will refuse to update model parameters if initial values are inconsistent with the observations[1]. The function used here z.init() initializes all unknown true states to be "alive" unless a harvest occurs. This works because fish not detected could either be alive or be a natural death, so "alive" is a valid initial value. Individuals that are ultimately harvested are known to be alive up to the period prior to harvest. It can be instructive to compare the matrices containing true and observed states with the initial values. For example, initial values can be saved by entering z.test <- z.init(y) in the Console. Entering sim$CH.TRUE[1:5,] and CH[1:5,] in the Console will display true and observed states for the first five individuals for comparison to z.test.

Results for the chosen simulation settings are moderately reliable. There are many parameters, as the model estimates true latent states as well as the probabilities for survival, harvest and natural death. It seems to consistently detect the increase in harvest rate in period 3. Uncertainty increases toward the end of the study, especially for probabilities of detection and natural death. This makes sense, because fish not detected on the final occasion could either be alive or be a natural death.

Perhaps the most obvious simulation settings to vary are the fairly robust sample size of tagged fish (100) and the detection probabilities (0.6 and above). Uncertainty increases markedly as detection probabilities decrease, so these simulations can be very helpful in planning the field study (e.g., number of

---

[1]https://www.vogelwarte.ch/de/projekte/publikationen/bpa/code-for-running-bpa-using-jags

receiver stations). The number of occasions can also be varied, but it requires commensurate changes in the vectors for u, v, and p.

7.5   Exercises

1. For the exploitation rate study design (Section 7.1), modify the JAGS code to plot the posterior distribution for the exploitation rate. Include a vertical line showing the underlying true exploitation rate.

2. For the completed tagging study design (Section 7.2), modify the code to use a single F parameter (representing an average over periods).

3. For the multi-year tagging study (Section 7.3), run three replicate simulations of the version using planted tags, with tag releases of 1000 (current code), 100, and 30. How does uncertainty (e.g., credible interval width) vary? What sample size would you recommend if planning this study? Also, save credible intervals for natural death rate for period 3 (v3) for the 1000 fish releases (for Exercise 4 below).

4. Compare results for the natural death rate in period 3 (v3) (from Exercise 3) with a modified design using five release and return periods. Extend the exploitation (u) and natural death (v) vectors by using the values from periods 1-2 for the final two periods. How do credible intervals for v3 change when the study is extended?

5. For the telemetry study (Section 7.4), compare results (particularly uncertainty) for the natural death rate for period 4 (v4) when the detection probability in period 5 (p5) takes on the following values: 0.2, 0.4, 0.7, 0.9. How might this affect planning for conducting the field work?

8

Growth

Growth is a basic aspect of a fish's life history, but it is also an important process to consider in managing fish populations. For example, a minimum size limit for harvest might be used to increase the proportion of fish surviving to the size (and age) of sexual maturity. Another consideration in developing a harvest strategy is finding a balance between the growth rate (increasing the size of each individual in a cohort) and the loss rate due to natural mortality (reducing the number of individuals in a cohort).

There are typically two aspects to growth modeling: age versus length and length versus weight. With those two equations, it is possible to predict the length or weight at any age, and to examine harvest strategies using minimum size or slot (legal harvest between lower and upper limits) regulations. The focus in commercial marine fisheries is typically on weight because income from commercial fishing is a function of weight of the catch. Length gets more emphasis in freshwater fisheries management because anglers are not selling their catches (and trophy designations are generally length-based).

8.1 Growth in length

The relationship between age and length for adult fish usually is a curve of decreasing slope; that is, growth gradually slows with age as individuals approach some average maximum size. The equation most often used for age and length is the von Bertalanffy growth function: $L_t = L_\infty * (1 - e^{-k*(t-t_0)})$. This equation for size at time 't' has three parameters: the asymptotic size or average maximum length (L_∞), growth rate (k), and time (age) at length of 0 (t_0). The last parameter is a theoretical one rather than something that is observed. It can take on positive or negative values, and can be thought of as an extrapolation from sizes at older ages.

In the following sections, we fit von Bertalanffy growth curves using two sources of data: paired age:length observations from a laboratory examination of otoliths or other hard parts that form annual marks (Campana, 2001), or a tagging study.

8.1.1   Age:length

Age:length observations may come from a survey or from sampling a fishery. Some ageing error invariably occurs when interpreting otoliths or other hard parts (Campana, 2001), but for simplicity I assume that the ages are determined without error. I further assume that fish are not sampled randomly but are instead a systematic sample by age. This ensures that the data set includes less common ages (sizes), particularly older fish.

The first step in fitting the model is to generate a simulated age:length data set.

```
# Fitting von Bertalanffy curve to simulated data
rm(list=ls()) # Clear Environment

# Choose von Bertalanffy parameter values and other simulation settings
L_inf <- 80   # e.g., 80-cm fish
k <- 0.3 # Slope
t_0 <- -0.5 # Age at length 0
MaxAge <- 12
SystSample <- 5   # Number per age group
N.AgeLen <- SystSample * MaxAge
Age <- rep(1:MaxAge, each=SystSample)
Var_L <- 10 # Variance about age:length relationship
Len <- rnorm(n=N.AgeLen,L_inf*(1-exp(-k*(Age-t_0))), sd=sqrt(Var_L))

plot(Age, Len, xlab="Age", ylab="Length (cm)")
```

The parameter values and other simulation settings provide a reasonable sample size and well-defined growth pattern and asymptote. The latter characteristic is critical in obtaining reliable results. For real data sets, it is always worthwhile to look at the shape of the scatterplot of length versus age. Data sets with few young fish or lacking a well-defined asymptote will often result in poor estimates. Our simulation uses integer ages but fractional (real-valued) ages could be used if there was good information about the timing of spawning and formation of annual marks on hard parts. Using fractional ages would be more important for short-lived species. The model fitting process is the same whether using integer or fractional ages.

We assume an additive error term; i.e., variability in length at age is constant. Whether an additive or multiplicative error term is appropriate can often be seen by examining a scatterplot of length versus age. A multiplicative error term would be called for if points have greater spread for older fish.

The JAGS code for fitting the model is generally similar to the simulation, except that the dnorm() function for the likelihood requires precision, defined as the inverse of the variance. Vague uniform priors are used for all four parameters. The prior distribution ranges and initial values may need to be adjusted if the range of simulated (or observed) lengths is varied. Note that max(Len) is used as a convenient upper bound for initial values of L_∞, but a fixed value greater than observed lengths could also be used.

```
# Load necessary library packages
library(rjags)
library(R2jags)

# JAGS code
sink("GrowthCurve.txt")
cat("
model{

# Priors

  L_inf.est ~ dunif(0, 200)
  k.est ~ dunif(0, 2)
  t0.est ~ dunif(-5, 5)
  Var_L.est ~ dunif(0,100)    # Variability in length at age

# Calculated value
  tau.est <- 1/Var_L.est

# Likelihood
  for (i in 1:N.AgeLen) {
     Len_hat[i] <- L_inf.est*(1-exp(-k.est*(Age[i]-t0.est)))
     Len[i] ~ dnorm(Len_hat[i], tau.est)
  } #i
}
    ",fill=TRUE)
sink()

# Bundle data
jags.data <- list("N.AgeLen", "Age", "Len")

# Initial values
jags.inits <- function(){list(L_inf.est=runif(n=1, min=0, max=max(Len)),
                              k.est=runif(n=1, min=0, max=2),
                              t0.est=runif(n=1, min=-5, max=5),
                              Var_L.est=runif(n=1, min=0, max=100))}

model.file <- 'GrowthCurve.txt'

# Parameters monitored
```

```
jags.params <- c("L_inf.est", "k.est", "t0.est", "Var_L.est")

# Call JAGS from R
jagsfit <- jags(data=jags.data, jags.params, inits=jags.inits,
                n.chains = 3, n.thin=1, n.iter = 10000,
                model.file)
print(jagsfit)
plot(jagsfit)
```

Convergence is generally rapid, and the estimates from simulated data are
reliable. Try varying some of the simulation settings and parameter values to see
how the model performs under different conditions. For example, note how the
growth curve changes as you try different values for the von Bertalanffy growth
parameters and the variance about the curve. One caution is to ensure that the
assumed additive error term does not produce negative length values. If that
occurs, one solution would be to switch to a multiplicative error term, using a
lognormal rather than a normal distribution; see Section 8.2. Another setting
to vary is the maximum age. A low value, as might occur if the population was
heavily exploited, can make it difficult to estimate L_∞. A Bayesian approach for
dealing with limited data is to use an informative prior distribution (Doll and
Jacquemin, 2018). This curve is particularly well suited to use of informative
priors, given the wealth of published information about fish growth and von
Bertalanffy parameters.

The uniform prior distributions chosen for this example were intended to
be uninformative, and additional testing with large sample sizes showed no
apparent bias. Lemoine (2019) argues against using priors with fixed limits and
instead recommends using an unbounded, positive-only distribution. Trying
multiple priors is a practical way to establish that the results are robust to
the particular distributions chosen. For example, similar results were obtained
when using a vague lognormal prior (e.g., dlnorm(0,1E-6)) with initial values
from rlnorm(n=1, meanlog=0, sdlog=0.5)) for L_∞, k, and Var_L, which are
restricted to positive values. A vague normal distribution (e.g., dnorm(0,1E-6)),
with initial values from rnorm(n=1, mean=0, sd=1)) was used for t_0, which can
take on positive or negative values. Note that the model structure can affect
whether a prior distribution is uninformative (McCarthy, 2007). For example,
the uniform prior for Var_L will not be flat when transformed to calculate
precision (tau). In general, a prior distribution chosen to be uninformative
will not be for transformed parameters. One way of examining this issue for
the growth model would be to fit the model using a prior for tau (so that
Var_L is the calculated value). Test runs with large sample sizes showed similar
results for the original model and when a vague uniform prior was used for
tau.

An empirical example (Quinn and Deriso, 1999) of fitting a von Bertalanffy growth curve to age and length data for rougheye rockfish (*Sebastes aleutianus*) is included as Appendix A.5.

8.1.2 Using tagging data

Growth information can also be obtained from a tagging study (Wang et al., 1995; Zhang et al., 2009; Scherrer et al., 2021). Fish seen on two or more occasions provide information on growth as a function of size at initial capture, time-at-large (interval between tagging and a subsequent recapture) and the growth increment. This approach differs from using age and length data because age at tagging is not known. However, it can be made compatible with a traditional age:length analysis by estimating relative age (true age - t_0) of all individuals as additional parameters (Wang et al., 1995; Zhang et al., 2009; Scherrer et al., 2021). This allows growth to be modeled as a function of age rather than size. Flexible versions of these models can allow for separate growth parameters (L_∞, k) for each individual. Here we fit a simpler model (Model 4 of Scherrer et al., 2021) with all individuals sharing L_∞ and k. It is straightforward to modify the code to allow for either or both parameters to vary among individuals (Scherrer et al., 2021).

One important practical question is how to obtain length data when fish are encountered after tagging. Using lengths from fishers provides the greatest sample size but runs the risk of substantial measurement error or outright bias, if an angler "estimates" the length. Length data obtained by biologists limits the sample size but ensures data quality, and could be obtained through field survey sampling or fishery monitoring by creel clerks or at-sea observers. In order for our results to apply to the whole population, we also assume that growth is the same for tagged and untagged fish. The model allows for measurement error and other sources of variability (e.g., environmental, genetic) that cause individual fish lengths to vary about the fitted curve. A constant additive error is assumed here, but the model could be modified to allow for a multiplicative error term. We begin with simulation code for generating the observations:

```
rm(list=ls()) # Clear Environment

L_inf <- 80  # e.g., 80-cm fish
k <- 0.3 # Slope
Var_L <- 10 # Variance about age:length relationship
N.Tag <- 120 # Number of fish recaptured at least once
n <- rpois(N.Tag, lambda=0.1)+2
    # Number of encounters for each recaptured individual, >2 uncommon

# Generate vector for relative age at initial capture (true age + t0)
```

```
Shape <- 5 # Parameters for gamma distribution generating relative ages
Rate <- 2
A <- rgamma(n=N.Tag, shape=Shape, rate=Rate) # Relative age vector
hist(A, xlab="Relative age", main="")

L <- array(data=NA, dim=c(N.Tag, max(n)))
dt <- array(data=NA, dim=c(N.Tag, max(n)-1))
for (i in 1:N.Tag){
  for (j in 1:(n[i]-1)){
    dt[i,j] <- rgamma(n=1, shape=6, rate=2)
      # Random dist of times at large
  }
}
hist(dt, xlab="Time at large", main="")

for (i in 1:N.Tag){
  L[i,1] <- rnorm(n=1, mean=L_inf *(1.0 - exp(-k*A[i])), sqrt(Var_L))
  for (j in 2:n[i]){
    L[i, j] <- rnorm(n=1, L_inf*(1.0 - exp(-k*(A[i]+dt[i, j-1]))),
                    sqrt(Var_L))
  } #j
} #i

plot(A, L[,1], xlab="Relative age", ylab="Length at first capture")
plot(dt[,1], (L[,2]-L[,1]), xlab="Time at large",
     ylab="Growth increment")
plot(L[,1], (L[,2]-L[,1]), xlab="Length at first capture",
     ylab="Growth increment")
```

Choices for the growth parameters are arbitrary but should produce a well-defined asymptote when combined with the other settings (time-at-large, relative age). The sample size of fish recaptured at least once is relatively robust but can of course be varied to judge model performance given fewer or more individuals. To allow for random variation in the tagging data set, a Poisson distribution (Section 3.1.4) is used to generate the number of encounters for each individual. Each individual is seen at least twice, but the lambda parameter can be adjusted to vary the typical number of encounters (use hist(n) to see distribution). As in previous studies (Wang et al., 1995; Zhang et al., 2009; Scherrer et al., 2021), we use a gamma distribution (Section 3.2.5) for relative age. The two gamma parameters (shape and rate) can be varied to obtain relative age distributions with desired characteristics (as observed on the hist() plot). Here they are chosen to provide a data set of mostly younger fish. Times-at-large are also generated using a gamma distribution, with shape and rate settings chosen to obtain a realistic pattern (mostly shorter times). Three plots are used to examine the tagging data set. The first (relative age versus length) may not have a well-defined asymptote if mostly younger fish are tagged. The plot can be examined to ensure that the chosen variance in length at age is reasonable (and not generating negative lengths). The second and third plots are noisy and difficult to interpret because growth increment is a function of both time-at-large and size-at-tagging. They are nevertheless useful for ensuring that the data set encompasses the early faster-growing period and slowed growth near the asymptote.

The JAGS code for fitting the model uses uninformative uniform prior distributions for the two growth parameters, variance in length-at-age, and the two gamma hyperparameters that describe the distribution of relative ages.

```
# Load necessary library packages
library(rjags)
library(R2jags)

# JAGS code
# Code modified from Scherrer et al. 2021: model 4
sink("vonBert_GrowthInc.txt")
cat("
model{
# Priors
  Var_L.est ~ dunif(0, 100) # Variance about expected length
  tau.est <- 1/Var_L.est # Precision
    k.est ~ dunif(0, 2)
    L_inf.est ~ dunif(0, 200)
    Shape.est ~ dunif(0, 100) # Gamma hyperparameters for relative age
    Rate.est ~ dunif(0, 100)

    # Likelihood
    for (i in 1:N.Tag)   {
        A.est[i] ~ dgamma(Shape.est, Rate.est)
          # Relative age (true age + t0)
        L_Exp[i, 1] <-   L_inf.est *(1.0 - exp(-k.est*A.est[i]))
          # Expected length at capture
        L[i, 1] ~ dnorm(L_Exp[i, 1], tau.est)
          # Length at initial capture
        for (j in 2:n[i])   { # Recapture lengths
            L_Exp[i, j] <-  L_inf.est*(1.0 -
                              exp(-k.est*(A.est[i]+dt[i, j-1])))
            L[i, j] ~ dnorm(L_Exp[i, j], tau.est)
          } #j
      } #i


}
```

```
    ",fill=TRUE)
sink()

# Bundle data
jags.data <- list("N.Tag", "n", "L", "dt")

# Initial values
jags.inits <- function(){ list(L_inf.est=runif(n=1, min=0, max=200),
                               k.est=runif(n=1, min=0, max=2),
                               Var_L.est=runif(n=1, min=0, max=100),
                               Shape.est=runif(n=1, min=0, max=100),
                               Rate.est=runif(n=1, min=0, max=100)
                               )}

model.file <- 'vonBert_GrowthInc.txt'

# Parameters monitored
jags.params <- c("L_inf.est", "k.est", "Var_L.est" ,"Shape.est",
                 "Rate.est")

# Call JAGS from R
jagsfit <- jags(data=jags.data, jags.params, inits=jags.inits,
                n.chains = 3, n.thin=1, n.iter = 20000,
                model.file)
print(jagsfit)
plot(jagsfit)
```

The model requires a considerable number of updates to achieve convergence, which is understandable given the large number of parameters (due to estimating relative age for each individual). The estimates tend to be reliable, which can be attributed to the large sample size and good ranges for length-at-tagging and time-at-large. Varying the simulation settings for time-at-large and relative age are helpful for understanding this approach to growth analysis. It should not matter for this model whether observations are from more individuals or more captures per individual. Increasing the number of observations per individual (Poisson lambda) would be relevant if fitting a model with growth parameters that varied among individuals.

For a given sample size, precision is lower for tagging compared to age:length data. This seems reasonable given that relative age must be estimated in a tagging study, whereas absolute ages are (assumed to be) known in an age:length analysis. Nevertheless, there are some situations where tagging data are preferred to age:length, such as for tropical fishes that do not show annual marks on hard parts.

It is straightforward to carry out a combined analysis using age:length and tagging data. The age:length data provide information on all three growth parameters (L_∞, k, t_0) as well as the variance term. The tagging data provide information on all parameters except t_0. A combined analysis might be preferable if multiple methods provide better coverage over the entire age/size range. For example, Scherrer et al. (2021) demonstrated the benefits of a combined analysis using age:length, tagging, and length frequency data.

8.2 Growth in weight

The length-weight relationship for fish is typically nonlinear: $W = a * L^b$. The 'a' parameter is a scaling coefficient and varies depending on the units chosen for measuring length and weight (e.g., cm versus mm or kg versus g). The 'b' parameter is close to 3 for most fishes, because weight is a three-dimensional value, increasing approximately as a cube of length.

The length-weight relationship is useful for several reasons. It provides information about fish condition (shape). Fish in poor condition may be overcrowded (e.g., in a pond or aquaculture setting) or lacking in suitable forage. A practical reason for developing a length-weight relationship is that length can be measured more quickly and accurately than weight, especially on the water. Fortunately, there tends to be a very strong relationship between length and weight, so it is often advantageous to measure length and predict weight.

We begin with simulation code, using an arbitrary range of 20 to 80 for length (e.g., measured in cm). For our example, the scaling parameter 'a' will be around 1E-6, so we instead work with the ln-scale value (-12). The model can be fitted using either 'a' or the ln-scale transformed parameter. Any assumed value around 3 is fine for the 'b' parameter. A multiplicative error term is assumed, as variability in weight typically increases as a function of length (Hayes and Brodziack, 1997).

```
# Fitting length:weight relationship to simulated data
rm(list=ls()) # Clear Environment

# Choose parameter values and other simulation settings
ln_L_W_a <- -12 # e.g., L in cm, weight in kg
L_W_b <- 3.2
MinL <- 20
MaxL <- 80
SystSample <- 2   # Number per length bin
```

```
Len <- rep(seq(from=MinL, to=MaxL, by=1), each=SystSample)
SampSize <- length(Len)
W_Var <- 0.1 # Ln-scale variance in weight at length
Wgt <- rlnorm(n=SampSize, mean=(ln_L_W_a+log(Len^(L_W_b))),
              sd=sqrt(W_Var))

par(mfcol=c(1,1)) # Ensure default of 1 panel (changes in trace plots)
par(mar=c(5.1, 4.1, 4.1, 2.1)) # Set to default margins
plot(Len, Wgt, xlab="Length (cm)", ylab="Weight (kg)")
```

The simulation uses a systematic sample for length, using the seq() function nested within rep(). Use Help for the rep() function and try each function separately in the Console to understand how the two functions work together to generate the length vector. The plot par() function is used to undo any plot formatting changes (from the trace plots generated below). As the plot illustrates, variability around the underlying length-weight relationship increases as a function of length.

The JAGS code is relatively short and fitting appears straightforward, although appearances can be deceiving! This simple model is probably the most difficult one to fit in this book, because the two parameters are highly correlated. The high correlation means that offsetting changes in 'a' and 'b' can provide essentially the same fit, so the updating process is very slow. This model provides a good opportunity to learn about the updating process and using Rhat and trace plots to judge convergence.

The JAGS code for fitting the model begins with broad uniform ranges for the prior distributions. Note that we use same lognormal distribution for the likelihood as in the simulation code (dlnorm() in JAGS). The lognormal distribution assumes a constant additive error term which is a multiplicative error in the back-transformed arithmetic scale. It is convenient to work with the ln-scale transformed value $(log(a) + log(L^b))$, but the model can also be fitted working directly with the 'a' parameter $(log(aL^b))$.

```
# Load necessary library packages
library(rjags)
library(R2jags)

# JAGS code
sink("LenWgt.txt")
cat("
model{
```

```
# Priors
 ln_L_W_a.est ~ dunif(-20, 20)
 L_W_b.est ~ dunif(1, 4)
 W_Var.est ~ dunif(0,0.5)
    # Multiplicative error. Ln-scale variability

# Calculated value
 precision <- 1/W_Var.est

# Likelihood
 for (i in 1:SampSize) {
    lnWgt_hat[i] <- ln_L_W_a.est+log(Len[i]^L_W_b.est)
    Wgt[i] ~ dlnorm(lnWgt_hat[i], precision)
 } #i
}
    ",fill=TRUE)
sink()

# Bundle data
jags.data <- list("SampSize", "Wgt", "Len")

# Initial values

jags.inits <- function(){ list(ln_L_W_a.est=runif(n=1, min=-20, 20),
                               L_W_b.est=runif(n=1, min=1, max=4),
                               W_Var.est=runif(n=1, min=0, max=0.5))}


model.file <- 'LenWgt.txt'

# Parameters monitored
jags.params <- c("ln_L_W_a.est", "L_W_b.est", "W_Var.est")

# Call JAGS from R
jagsfit <- jags(data=jags.data, jags.params, inits=jags.inits,
                n.chains = 3, n.thin=1, n.iter = 5000,
                model.file)
print(jagsfit)
plot(jagsfit)

par(mar=c(5.1, 4.1, 4.1, 2.1)) # Set default margins
plot(jagsfit$BUGSoutput$sims.list$ln_L_W_a.est,
     jagsfit$BUGSoutput$sims.list$L_W_b.est)
cor(jagsfit$BUGSoutput$sims.list$ln_L_W_a.est,
    jagsfit$BUGSoutput$sims.list$L_W_b.est)
```

```
jagsfit.mcmc <- as.mcmc(jagsfit) # Creates an MCMC object for plotting
par(mar=c(1,1,1,1)) # Avoid error for "figure margins too large"
plot(jagsfit.mcmc) # Trace and density plots
autocorr.diag(jagsfit.mcmc)
```

With 5,000 MCMC iterations, the estimates are generally close to the true
values, but in some runs Rhat values are inflated. Plotting the sequence of
MCMC estimates for ln-scale 'a' versus 'b' makes clear the extremely high
correlation between the two parameters. A numerical estimate of the correlation
(~0.99) is obtained in the Console window by using the cor() function.

The trace plots show poor mixing for the 'a' and 'b' parameters and sometimes
more than one mode in the density plots, indicating that many additional
updates are needed (Figure 8.1). Compare this result to a longer run using
50,000 updates. The trace plots for 'a' and 'b' now have a much better
appearance, and we have good low estimates for Rhat. The autocorrelations
can be reduced by thinning (e.g., try n.thin=10), which also reduces the
memory required.

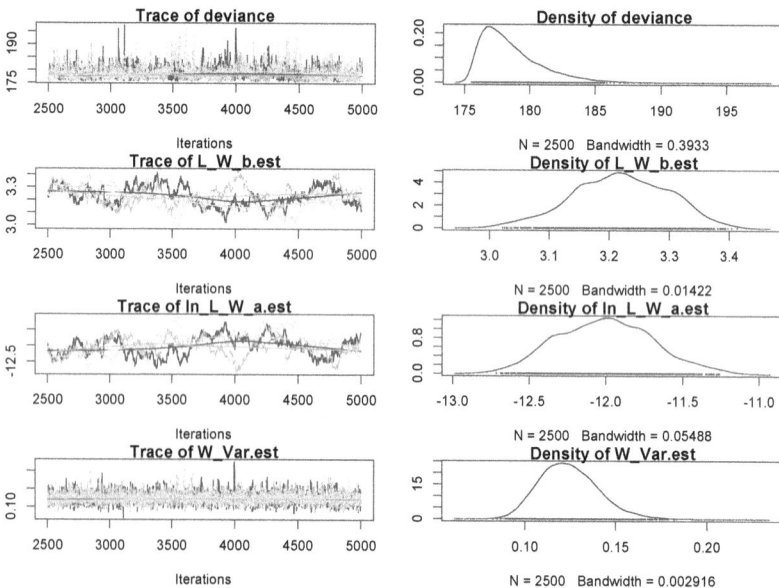

FIGURE 8.1 Example trace and density plots for length-weight relationship,
based on 5000 updates.

It can be helpful to add code for displaying the length-weight observations and
the fitted curve. The as.numeric() function converts each array (of length 1)

to a scalar for use in generating the fitted curve, which is added to the plot
using the points function.

```
par(mar=c(5.1, 4.1, 4.1, 2.1)) # Restore default margins
plot(Len, Wgt, xlab="Length (cm)", ylab="Weight (kg)")
par_a <- exp(as.numeric(jagsfit$BUGSoutput$median$ln_L_W_a.est))
par_b <- as.numeric(jagsfit$BUGSoutput$median$L_W_b.est)
Wgt_hat <- par_a*(Len^par_b)
points(Len, Wgt_hat, type="l", lty=1)
```

8.3 Exercises

1. Before carrying out simulation runs (code from Section 8.1.1), describe the expected outcome in fitting the von Bertalanffy growth
curve to age:length data if you changed the growth rate (k) to 0.05.
Report the results for asymptotic size estimates from several runs.
What other simulation setting would need to change in order to
accommodate such a slow growth rate?

2. For the model in Section 8.1.1, compare estimates from the default
model (vague uniform prior distributions) to a model using vague
lognormal priors for L_∞, growth rate (k), and Var_L, and a vague
normal prior for t_0.

3. For either the default or lognormal case from Exercise 2, modify the
code to add the fitted curve to the age:length scatterplot. As a hint,
you can either redo the scatterplot after the plot(jagsfit) line or add
a # to turn off the jagsfit plot.

4. Before carrying out simulation runs, describe the expected outcome
in fitting the von Bertalanffy growth curve to age:length data if
you changed the maximum age to 4 (e.g., if the stock was heavily
exploited). Report the results for asymptotic size estimates from
several runs.

5. Modify the code for the von Bertalanffy age:length analysis to obtain
the Age vector using the rgamma function that provided relative
ages for the tagging analysis. Find a pair of values for shape and
rate that result in reliable estimates of the growth parameters.

9

Recruitment

Growth, mortality, and recruitment are the three fundamental processes that regulate fish populations. Previous chapters have addressed the first two; here we consider methods for studying recruitment. Recruitment is usually considered to be the annual process of young fish entering the part of the population that is vulnerable to fishing (i.e., the fishable stock). Thus the timing of recruitment depends on fishery gear selectivity pattern(s), which tend to be gradual rather than knife-edged. One practical approach for defining age at recruitment is to choose the first age at which fishery catches are non-negligible. For simplicity in this chapter, we will simulate recruitment occurring at age 1, so there is a one-year lag from the time of spawning to recruitment.

What happens between spawning and recruitment is of interest to the fish ecologist and the fishery manager. Typical factors affecting recruitment or year-class strength include environmental conditions, prey availability, and predation. It may seem obvious to assume that spawning stock size affects recruitment, but it is not always easy to detect such a relationship (Hilborn and Walters, 1992). The relationship may not be evident if there is little contrast in spawning stock size, high variability in recruitment, too short a time series, or low precision for estimates of spawning stock or recruitment. Characterizing the strength of the relationship can provide valuable guidance to the fishery manager. For example, management will need to be more restrictive in cases where recruitment drops consistently as spawning stock declines. If recruitment appears to be relatively stable over a wide range of spawning stock sizes, a more aggressive harvest policy may be possible. The risk in any case is of fishing the stock down to a point where recruitment declines sharply. This is termed "recruitment overfishing", and it can be very difficult to rebuild a population after that occurs.

The process of recruitment starts with eggs for most fishes, but egg production is difficult to estimate directly so it is common to use spawning stock biomass (B_S) as a proxy. One of the most commonly used curves relating spawning stock size and recruitment is the Beverton-Holt model (Beverton and Holt, 1993):

$$R = 1/(\alpha + \beta/B_S)$$

This curve approaches zero as B_S approaches zero and approaches $1/\alpha$ as B_S approaches infinity. It is often a good descriptor of observed stock-recruitment data, in that recruitment is low when spawning stock size is close to zero but appears to be without trend at higher spawning stock sizes.

The best way to explore the behavior of the Beverton-Holt model is to try different values for the two parameters:

```
SpBio <- seq(from=10, to=200, by=5)
a <- 0.001
b <- 0.1
R <- 1/(a+b/SpBio)
plot(SpBio, R)
```

In this example code, spawning stock biomass is generated with the seq() (sequence) function over a wide but arbitrary range. The Beverton-Holt parameters are also arbitrary but are chosen to allow for gradually increasing recruitment toward an asymptote of 1000 (1/a). Experiment with different parameter values to find a set that would maintain an asymptote of 1000 but would produce a steeper curve (near the origin), with about 90% of asymptotic recruitment at a spawning stock level of 100.

9.1   Fitting a stock-recruitment curve

Our simulation will use a full age-structured model to produce the spawning stock and recruitment observations for fitting a Beverton-Holt model. First, we choose the structure of the population matrix (number of ages and years). Next, we choose arbitrary annual fishing mortality rates. By starting the population at a high level and fishing intensively, we ensure that there is sufficient contrast (Hilborn and Walters, 1992) in spawning stock size to be able to detect the underlying relationship. Age-specific fishing mortality rates are obtained by defining the fishery gear selectivity pattern. We use a logistic curve and arbitrarily choose parameters so that 50% selectivity occurs at age 2. (It is useful to plot the selectivity curve (code commented out) while varying the two selectivity parameters.) Obtaining year- and age-specific fishing mortality rates as a product of a year-specific measure of fishing intensity and an age-specific vulnerability to fishing is sometimes referred to as a "separable" model.

```
rm(list=ls()) # Clear Environment
```

```
A <- 10 # Arbitrary. If changed, redo maturity and meanwt vectors
Y <- 12 # Arbitrary. If changed, redo AnnualF vector

AnnualF <- c(0.59, 0.41, 0.74, 0.91, 0.86, 0.74, 1.07, 0.9, 0.87,
             1.1, 0.93)
# Arbitrary, increasing trend to create contrast in SpBio.
# Redo if Y changes
AnnualM <- 0.2

# F by age, using separable model
Fishery_k <- 1 # Slope for logistic function
Fishery_a_50 <- 2 # Age at 0.5 selectivity
SelF <- array(data=NA, dim=A) # Fishery selectivity pattern
F_a <- array(data=NA, dim=c(A,(Y-1))) # F by age and year
Z_a <- array(data=NA, dim=c(A,(Y-1))) # Z by age and year
S_a <- array(data=NA, dim=c(A,(Y-1))) # Survival rate by age and year
for (a in 1:A){
  SelF[a] <- 1/(1+exp(-Fishery_k*(a-Fishery_a_50)))
  } #a
#plot(SelF)
for (y in 1:(Y-1)) {
  for (a in 1:A){
    F_a[a,y] <- AnnualF[y] * SelF[a]
    Z_a[a,y] <- F_a[a,y]+AnnualM
    S_a[a,y] <- exp(-Z_a[a,y])
    } #a
} #y

N <- array(data=NA, dim=c(A, Y))

BH_alpha <- 1/1E5 # Beverton-Holt asymptotic recruitment
BH_beta <- 0.1 # Arbitrary
V_rec <- 0.1 # Log-scale recruitment variance
SD_rec <- sqrt(V_rec)

# Set up year-1 vector based on asymptotic recruitment
# and year-1 survival
Mat <- c(0, 0, 0.2, 0.4, 0.8, 1, 1, 1, 1, 1)
# Arbitrary maturity schedule. Redo if max age (A) changes
Wgt <- c(0.01, 0.05, 0.10, 0.20, 0.40, 0.62, 0.80, 1.01, 1.30, 1.56)
# Mean wt. Redo if A changes
SpBio <- array(data=NA, dim=(Y-1)) # Spawning biomass vector
# Year-1 N (arbitrarily) from asymptotic recruitment, year-1 survival.
N[1,1] <- trunc(1/BH_alpha*rlnorm(n=1, 0, SD_rec))
```

```
for (a in 2:A){
  N[a,1] <- rbinom(n=1, N[(a-1),1], S_a[(a-1),1])
}#a
for (y in 2:Y){
  SpBio[y-1] <- sum(N[,y-1]*Wgt*Mat) # Spawning stock biomass
  N[1,y] <- trunc(1/(BH_alpha+BH_beta/SpBio[y-1])
                  *rlnorm(n=1, 0, SD_rec))
  for (a in 2:A){
    N[a,y] <- rbinom(1, N[(a-1),(y-1)], S_a[(a-1),(y-1)])
  } #a
} #y
#plot(SpBio,N[1,2:Y], ylab="Recruits", xlab="Spawning stock biomass")

# Generate survey data on recruitment and spawning stock
Survey_q <- 1E-3 # Arbitrary catchability coefficient for survey
SD_survey <- 0.2 # Arbitrary ln-scale SD
Exp_ln_B <- log(Survey_q*SpBio)
Survey_B <- rlnorm(n=Y, meanlog=Exp_ln_B, sdlog=SD_survey)
#plot(SpBio, Survey_B[1:Y-1])
Exp_ln_R <- log(Survey_q*N[1,2:Y])
Survey_R <- c(NA,rlnorm(n=(Y-1), meanlog=Exp_ln_R, sdlog=SD_survey))
#plot(N[1,], Survey_R)
```

The next section of code generates the population matrix. The Beverton-Holt parameters are arbitrarily chosen and can be varied to explore different shapes for the curve. Next, we generate the population vector for year 1, using approximate equilibrium values. Expected recruitment is set equal to the curve's asymptote, with a lognormal (Section 3.2.4) variance (default 0.1) to add random variation as might be due to environmental conditions. The lognormal distribution is well suited to recruitment data because of the potential for occasional extreme values (Hilborn and Walters, 1992). The rest of the year-1 vector is obtained from age-1 recruitment using year-1 mortality rates $(N_{a,1} = S_{a-1,1} * N_{a-1,1})$. The rbinom() function introduces stochastic variation in survival. Starting (age-1) population size is truncated because the binomial function requires a whole number for the size of the experiment.

The population vector for the remaining years can be filled in once the year-1 vector is determined. For example, recruitment in year 2 depends on spawning stock biomass in year 1. The age-specific vectors for maturity schedule and weight-at-age are arbitrary. Numbers of fish age 2 and older are obtained as survivors from the prior year's vector; i.e., $N_{a,y} = S_{a-1,y-1} * N_{a-1,y-1}$, with random variation in survival from a binomial distribution. The final step is to obtain survey observations for recruitment and spawning stock biomass. The random survey values are drawn from a lognormal distribution, using an arbitrary catchability coefficient and ln-scale standard deviation for error. (There is not a recruitment survey value for year 1 because the simulated recruitment was not based on spawning stock size.) The survey values are referred to as indices but could be estimates of the absolute magnitude of recruitment and spawning stock size. Estimates of absolute recruitment and spawning stock size are typically obtained from a stock assessment model, similar to the catch-at-age analysis illustrated in Section 10.2. Brooks and Deroba (2015) recommend including the stock-recruitment relationship as an integral part of the catch-at-age analysis to avoid some issues with treating the stock assessment results as data.

The JAGS code uses uninformative prior distributions for the two Beverton-Holt parameters and the lognormal error term. Note that the JAGS function dlnorm() uses the inverse of the variance, whereas the R function for generating lognormal random variates rlnorm() uses the standard deviation. Expected recruitment is predicted on ln-scale in order to match the assumed lognormal error. The recruitment index is predicted (R_hat) for use in plotting (see below) and the DIC score is saved for use in Section 9.2.

```
# Load necessary library packages
library(rjags)
library(R2jags)

sink("StockRecruit.txt")
cat("
model {
# Model parameters: Beverton-Holt alpha and beta,
# ln-scale variance for fitted curve

# Priors
 BH_alpha.est ~ dunif(0, 10)
 BH_beta.est ~ dunif(0, 10)
 V_rec.est ~ dunif(0, 10)
 tau <- 1/V_rec.est

# Likelihood
    for(y in 2:Y) {
       lnR_hat[y] <- log(1/(BH_alpha.est+BH_beta.est/Survey_B[y-1]))
       Survey_R[y] ~ dlnorm(lnR_hat[y],tau)
       R_hat[y] <- exp(lnR_hat[y])
       } #y
}
    ",fill=TRUE)
sink()
```

```r
# Bundle data
jags.data <- list("Y", "Survey_R", "Survey_B")

# Initial values
y <- Survey_R[2:Y]
x <- Survey_B[1:(Y-1)]
nls_init <- nls(y ~ 1/(BH_a_start+BH_b_start/x),
                start = list(BH_a_start = 0.001, BH_b_start = 0.001),
                algorithm="port",
                lower=c(1E-6, 0), upper=c(1000,1000))
#summary(nls_init)
nls_save <- coef(nls_init)
#y_hat <- 1/(nls_save[1]+nls_save[2]/x)
#plot(x,y)
#points(x,y_hat, col="red", type="l")

jags.inits <- function(){ list(BH_alpha.est=nls_save[1],
                               BH_beta.est=nls_save[2],
                               V_rec.est=runif(n=1, min=0, max=10))}

model.file <- 'StockRecruit.txt'

# Parameters monitored
jags.params <- c("BH_alpha.est", "BH_beta.est", "V_rec.est",
                 "R_hat")

# Call JAGS from R
jagsfit <- jags(data=jags.data, jags.params, inits=jags.inits,
                n.chains = 3, n.thin=1, n.iter = 10000,
                model.file)
print(jagsfit)
plot(jagsfit)

BH.DIC <- jagsfit$BUGSoutput$DIC # Used in next section

df.xy <- data.frame(Survey_B[1:(Y-1)], Survey_R[2:Y],
                    jagsfit$BUGSoutput$median$R_hat)
df.xy <- df.xy[order(df.xy[,1]),]  # Sort by survey index for plotting
y.bounds <- range(Survey_R[2:Y],jagsfit$BUGSoutput$median$R_hat)

plot(df.xy[,1],df.xy[,2], ylab="Recruitment index",
     xlab="Spawning Stock Index", ylim=y.bounds) #c(min.y, max.y))
points(df.xy[,1], df.xy[,3], type="l")
```

Randomly chosen starting values for the two Beverton-Holt parameters sometimes caused JAGS to crash, so nls(), the R function for nonlinear least-squares, was used to generate good initial values. When unconstrained, nls() occasionally produced a negative estimate of the first parameter (1/asymptotic recruitment), so the "port" algorithm was used to allow bounds to be specified. The nls() code illustrates the compactness of using R functions, compared to a Bayesian analysis, but also the "black box" nature of the analysis.

The JAGS results tend to be reliable for the default settings. The plotted estimates of the recruitment indices capture the decrease due to declining spawning stock size. Estimates of the Beverton-Holt alpha parameter are typically close to 0.01 (BH_alpha/Survey_q), because we fit the curve to a recruitment index (true recruitment * Survey_q) rather than to absolute levels. Estimates of the Beverton-Holt beta parameter tend to be close to the true value. Estimates of the ln-scale variance about the curve tend to be higher than the true value, because the points vary due not only to true recruitment variation but also variation due to survey sampling.

As always, it is very helpful to view the points and fitted curve (Figure 9.1). We use a data frame to sort the spawning stock, recruitment, and predicted recruitment indices, in order to have a smooth-fitted curve (i.e., continuously increasing spawning stock values). (A data frame is a collection of variables, similar to a spreadsheet. Columns have the same number of observations but can be of different types.) Try multiple runs to see how variation in recruitment affects the pattern of points and fitted curve.

An empirical example of fitting a Beverton-Holt curve to data for North Sea plaice is included as Appendix A.6.

9.2 Does spawning stock size affect recruitment?

Section 9.1 illustrated the process of model fitting but did not answer the question of whether there is a detectable relationship between the spawning stock and recruitment indices. To address that question, we return to the approach shown in Section 4.3.7, i.e., comparing DIC scores for candidate models. Here the two candidates are the full Beverton-Holt model and a reduced model with constant recruitment.

We append JAGS code for the constant recruitment case at the end of the previous code so that both models are applied to the same spawning stock and recruitment data. Convergence is extremely rapid for this simpler case. The difference in DIC scores between the full (Beverton-Holt) and reduced

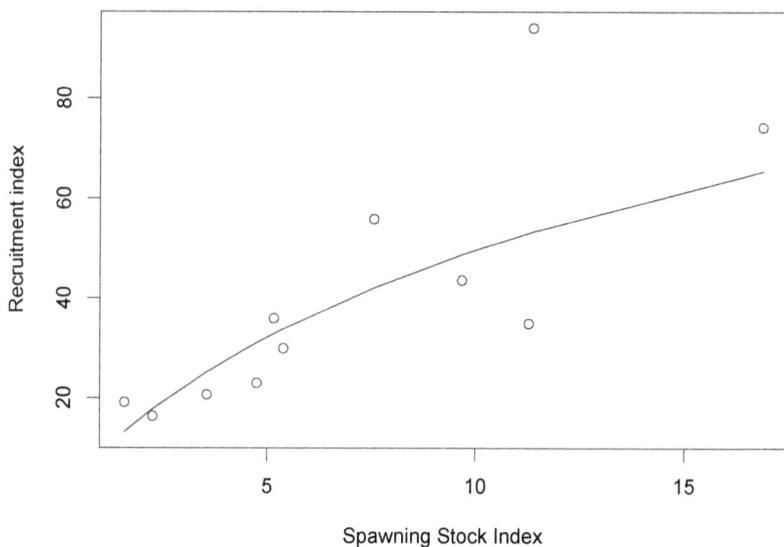

FIGURE 9.1 Example of fitted Beverton-Holt curve relating indices of spawning stock and recruitment.

(constant recruitment) models is printed to the Console, but it is also useful to compare pV, deviance, and DIC scores.

```
# Constant recruitment fit
# JAGS code ################################

sink("StockRecruit.txt")
cat("
model {
# Model parameters: Beverton-Holt alpha, SD for fitted curve

# Priors
 BH_alpha.est ~ dunif(0, 10)
 V_rec.est ~ dunif(0, 10)
 tau <- 1/V_rec.est

# Likelihood
    for(y in 2:Y) {
      lnR_hat[y] <- log(1/BH_alpha.est)
      Survey_R[y] ~ dlnorm(lnR_hat[y],tau)
      R_hat[y] <- exp(lnR_hat[y])
      } #y
```

```
    }
        ",fill=TRUE)
    sink()

    # Bundle data
    jags.data <- list("Y", "Survey_R")

    jags.inits <- function(){ list(BH_alpha.est=nls_save[1],
                                    V_rec.est=runif(n=1, min=0, max=10))}

    model.file <- 'StockRecruit.txt'

    # Parameters monitored
    jags.params <- c("BH_alpha.est", "V_rec.est", "R_hat")

    # Call JAGS from R
    jagsfit <- jags(data=jags.data, jags.params, inits=jags.inits,
                    n.chains = 3, n.thin=1, n.iter = 10000,
                    model.file)
    print(jagsfit)
    plot(jagsfit)
    ConstR.DIC <- jagsfit$BUGSoutput$DIC
    ConstR.DIC - BH.DIC
```

With the default settings, the DIC score is typically higher (indicating a poorer fit) for the constant recruitment model. Next, try varying some of the simulation settings. The pattern for annual fishing mortality could be altered to produce less of a decline over time in spawning stock (less contrast). Higher levels of recruitment variation will make the relationship harder to detect, and the DIC score may actually be lower for the constant recruitment model because of a smaller penalty for model complexity (smaller estimate of pV). Other potential settings to vary include the length of the time series and the precision of the survey data. Be sure to look back at the scatterplot and fitted Beverton-Holt curve, to get a sense of how the constant recruitment model would compare.

9.3 Exercises

1. Fit the Beverton-Holt model (Section 9.1) using the following values for the F array: runif(n=Y, min=0, max=0.1), runif(n=Y, min=0, max=1.0), and runif(n=Y, min=1.0, max=1.4). Present results from

five trials and provide an assessment of the reliability of estimates for each case.

2. The expected value for α (0.01 for our default simulation settings) is the true underlying value divided by the survey catchability, because we are fitting the model using survey indices rather than estimates of absolute abundance. Compare estimates of α for the Beverton-Holt and constant recruitment models, using the approach from Section 9.2. Explain any pattern that you detect.

10

Integrated Population Models

The field and analytical methods discussed thus far have addressed population rates one at a time (e.g., estimating abundance or survival). An improved approach would be to design a complementary set of field studies that provides a complete picture of population dynamics. For example, Conn et al. (2008) used harvest and tag recovery data to fit an integrated population model (IPM) for black bear (*Ursus americanus*). Their model provided estimates that were more precise and biologically realistic than point estimates using capture-recapture data alone. Kéry and Schaub (2012) developed an avian IPM using adult population counts, nest surveys, and capture-recapture data on survival. We begin here with a somewhat simpler approach, a two-age model (ages 1 and 2+) using capture-recapture and telemetry data, then examine a full age-structured model as typically used in fish stock assessment (Section 10.2).

10.1   Two-age model

This example shows how two independent field studies (capture-recapture estimates of abundance and telemetry information on survival) can be combined to make inferences about population dynamics. The cost of this approach in time and money is higher than a basic relative-abundance survey, but it provides insights about population dynamics (recruitment and survival) that would not be possible with relative abundance data alone. The cost could be justified for a subset of cases; for example, a population at low abundance or one supporting an important fishery. We begin with the simulation model:

```
rm(list=ls()) # Clear Environment

# Simulation to create population size and field data
Y <- 10
```

```
LowS <- 0.5 # Adult survival, arbitrary lower bound
HighS <- 0.9 # Arbitrary upper bound
AnnualS <- runif(n=(Y-1), min=LowS, max=HighS)

MeanR <- 400 # Arbitrary mean annual recruitment
N.r <- array(data=NA, dim=Y)
V_rec <- 0.1 # Log-scale recruitment variance
SD_rec <- sqrt(V_rec)
N.r <- trunc(MeanR * rlnorm(n=Y, 0, SD_rec))
N.a <- array(data=NA, dim=Y)
N.a[1] <- N.r[1] + rpois(n=1, lambda=MeanR)
  # Arbitrary year-1 population size
for (y in 2:Y){
  N.a[y] <- N.r[y] + rbinom(n=1, size=N.a[y-1], prob=AnnualS[y-1])
} #y

# Short-term (closed) two-sample population estimate
CapProb <- 0.2 # Fraction of population sampled
n1 <- rbinom(n=Y, size=N.a, prob=CapProb)
  # Caught and marked in first sample
n2 <- rbinom(n=Y, size=N.a, prob=CapProb)
  # Caught and examined for marks in second sample
m2 <- rbinom(n=Y, size=n2, prob=n1/N.a) # Marked fish in second sample

# Telemetry information on annual survival
TelRelease <- 20 # Number of telemetry tags to release annually
TelTags <- array(data=NA, dim=c((Y-1),Y))
for (y in 1:(Y-1)){
  TelTags[y,y] <- TelRelease
  for (t in ((y+1):Y)){
    TelTags[y,t] <- rbinom(n=1, size=TelTags[y,(t-1)],
                       prob=AnnualS[t-1])
  }} # y,t
```

We model absolute population size and assume that fish age 1 and older have the same survival rate, which varies annually within user-specified lower and upper bounds. Recruitment (surviving age-0 fish from the prior year) varies annually about an arbitrary mean according to a lognormal distribution (Section 9.1). We require a starting population in year 1 and draw the adult survivors from a Poisson distribution, using the same mean as recruitment. Population size in subsequent years is

$$N.a_y = N.r_y + Binomial(N.a_{y-1}, AnnualS_{y-1})$$

i.e., incoming recruitment plus surviving adults.

The next sections of code provide simulated data from two field studies that are assumed to be independent and would provide information on abundance and survival. The first data source is a two-sample (closed) capture-recapture estimate, assumed to be done at the start of each year. The capture probability, used to generate the two binomally distributed samples, is set at a relatively high default value to obtain better recruitment estimates. The number of recaptures is also binomally distributed, using the observed marked fraction (on average equal to the capture probability). The second data source is a telemetry study that provides annual information on the number of surviving telemetered fish. For example, fixed receivers might be used to detect migratory fish that return annually to a spawning area. We arbitrarily assume that 20 telemetered fish are released each year. For simplicity, transmitters are assumed to last for the duration of the study. Examine the TelTags matrix to see how the annual releases maintain a good sample size.

A hierarchical structure is used to estimate annual recruitment. The overall ln-scale mean and precision are estimated using uninformative priors. Annual recruitment estimates for years 2-Y are drawn from a lognormal distribution using the shared ln-scale mean and precision. (Recruitment in year 1 is used in the simulation but is not estimated because the population projection equation is only used in years 2-Y.) A uniform 0-1 prior is used for annual survival.

Population size in year 1 is Poisson distributed, using the capture-recapture point estimate as the expected value. Population size in subsequent years is also drawn from a Poisson distribution, using the population projection equation as the expected value. Both likelihoods use a binomial distribution (Section 3.1.2). The probability for the capture-recapture data is the estimated fraction of the population that is marked. The probability for the telemetry data is the estimated annual survival rate.

Following the JAGS section is some R code to compare the true values with model estimates. The par(xpd=TRUE) turns off clipping so that a legend can be added outside the plot boundary (using inset to set the relative coordinates of the legend). For the third plot, the Chapman modification of the capture-recapture estimator (Ricker, 1975) was used to avoid getting a plotting error due to an infinite point estimate in cases with $m_2 = 0$.

```
# Load necessary library packages
library(rjags)
library(R2jags)

# JAGS code ###############################
sink("IPM.txt")
cat("
```

```
model {

# Priors

 Mean_lnR.est ~ dnorm(0, 1E-6)   # Recruitment uses ln-scale parameter
 V_rec.est ~ dunif(0, 10)
 tau <- 1/V_rec.est
 for(y in 1:(Y-1)) {
    N.r.est[y] ~ dlnorm(Mean_lnR.est,tau)
     # Hierarchical structure for recruitment
  } #y

for (y in 1:(Y-1)){
  AnnualS.est[y] ~ dunif(0,1)   # Uninformative prior, adult survival
}#y

# Population projection
 N.a.est[1] ~ dpois((n1[1]+1)*(n2[1]+1)/(m2[1]+1)-1)
  # Year-1 total, Chapman mod in case m2=0
  # Needed to start model because no prior year information
 for(y in 2:Y) {
    N.a.est[y] ~ dpois(N.r.est[y-1] + (AnnualS.est[y-1]*N.a.est[y-1]))
     # Index shifted by one year for recruitment estimates
    } #y

#Likelihoods
# Capture-recapture estimate for total pop size at start of year
  for (y in 2:Y){   # m2[1] used above to get starting pop size
    m2[y] ~ dbin((n1[y]/N.a.est[y]), n2[y])
    } #y

# Telemetry information on survival rate
for (y in 1:(Y-1)){
  for (t in ((y+1):Y)){
    TelTags[y,t] ~ dbin(AnnualS.est[t-1],TelTags[y,(t-1)])
  }} # y, t
}
    ",fill=TRUE)
sink()

# Bundle data
jags.data <- list("Y", "n1", "n2", "m2", "TelTags")

# Initial values
jags.inits <- function(){ list(N.r.est= rlnorm(n=(Y-1), meanlog=10,
                                                 sdlog=1),
                          AnnualS.est=runif(n=(Y-1), min=0,
                                              max=1))}
```

```
model.file <- 'IPM.txt'

# Parameters monitored
jags.params <- c("AnnualS.est", "N.a.est", "N.r.est", "V_rec.est")

# Call JAGS from R
jagsfit <- jags(data=jags.data, jags.params, inits=jags.inits,
                n.chains = 3, n.thin=1, n.iter = 40000,
                model.file)
print(jagsfit)
plot(jagsfit)

# Code for plots comparing true values and model estimates
par(mfrow = c(1, 3))
plot(seq(1:(Y-1)),
     jagsfit$BUGSoutput$median$AnnualS.est, ylab="Annual Survival",
     xlab="Year", type="b", pch=19,
     ylim=range(AnnualS,jagsfit$BUGSoutput$median$AnnualS.est))
points(AnnualS, lty=3, type="b", pch=17)

plot(seq(1:(Y-1)),
     jagsfit$BUGSoutput$median$N.r.est,
     ylab="Annual recruitment (years 2-Y)", xlab="",
     type="b", pch=19,
     ylim=range(N.r[2:Y],jagsfit$BUGSoutput$median$N.r.est))
points(N.r[2:Y], lty=3, type="b", pch=17)
par(xpd=TRUE) # Turn off clipping to put legend above plot
legend("top", legend = c("Est", "True"), lty = c(1, 3),
       pch = c(19,17), text.col = "black",  horiz = T ,
       inset = c(0, -0.15))
par(xpd=FALSE)

MR_N.hat <- (n1+1)*(n2+1)/(m2+1) -1 # Chapman mod. in case m2=0
plot(seq(1:Y), jagsfit$BUGSoutput$median$N.a.est,
     ylab="Pop size (includes new recruits)", xlab="Year",
     type="b", pch=19,
     ylim=range(N.a,jagsfit$BUGSoutput$median$N.a.est,MR_N.hat))
points(N.a, lty=3, type="b", pch=17) # True adult pop size
points(MR_N.hat, lty=2, pch=0, type="b") # Point estimates
```

Results for the default case show that total population size and annual survival tend to be reliably estimated (Figure 10.1). Especially at lower capture probabilities, the capture-recapture point estimates (dashed line, open squares) tend to be more variable than the population estimates obtained from the integrated model (solid line). Annual recruitment tends to be less well estimated than total population size or survival rate. There is no direct information about

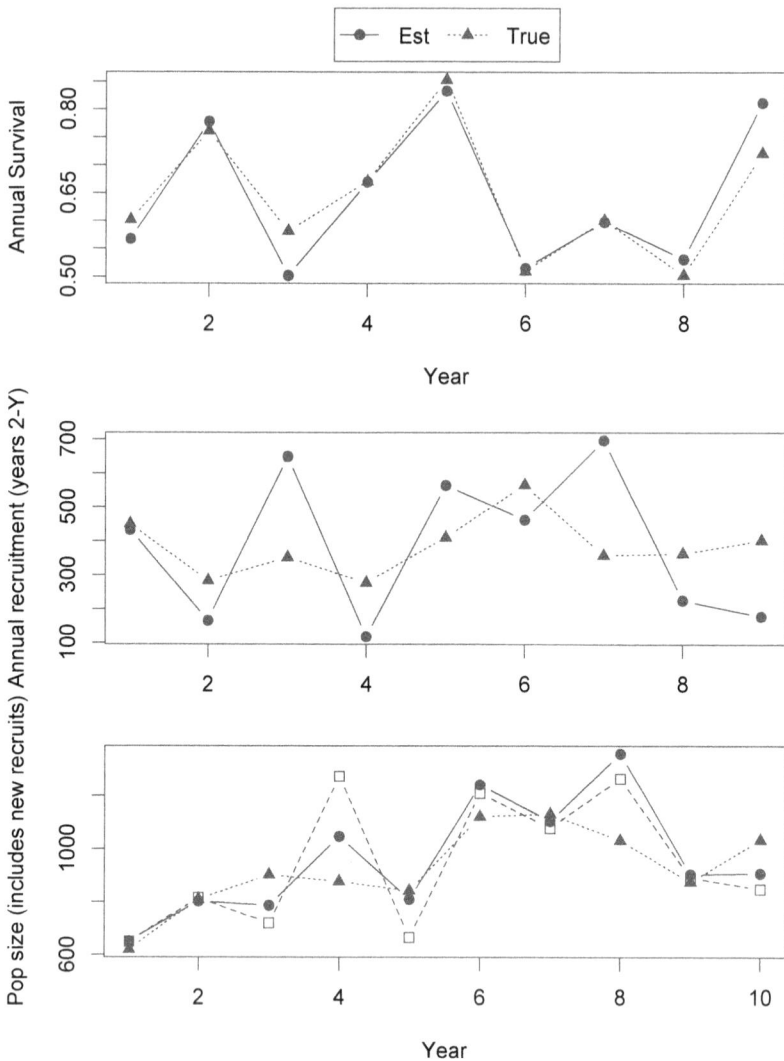

FIGURE 10.1 Estimates of annual survival (top), recruitment (middle), and total population size (bottom) from a two-age integrated population model fitted to capture-recapture and telemetry data. True values are shown with dotted lines; capture-recapture point estimates (bottom) are shown with a dashed line and open squares.

recruitment, so the model makes indirect inferences from total abundance (new age-1 recruits plus 2+ survivors) and telemetry information on survival rate.

After doing multiple runs with the default settings, try varying the capture probability for the capture-recapture study and the annual release of fish with transmitters. As expected, a higher capture probability improves the estimates of total population size, and more transmitters improve the annual survival estimates.

10.2 Full age-structured model

Another type of IPM is a fish stock assessment that typically uses catch by age and year to estimate population size and fishing mortality rate (Millar and Meyer, 2000; Schaub et al., 2024). Fish stock assessment models share many characteristics with IPMs developed for terrestrial species, but typically have a much greater spatial scale and are more focused on harvest regulation (Schaub et al., 2024). These models serve as the foundation for much of marine fishery management, as well as in freshwater fisheries where sufficient data are available (e.g., Fielder and Bence, 2014). Catch data are a valuable source of information because they reflect absolute abundance. Summing the catches from a cohort (year class) across years (along a diagonal in the catch-at-age matrix) provides a conservative estimate of the initial abundance of that cohort. Early versions of catch-at-age analysis (VPA; cohort analysis) were simple upward adjustments of the summed catches in order to account for natural mortality (Ricker, 1975). Current catch-at-age analyses have a statistical foundation using either maximum-likelihood or Bayesian estimation (Maunder and Punt, 2013). Catch data alone are insufficient to obtain reliable results, so current assessments integrate fishery catch data with auxiliary information from the fishery (e.g., annual level of fishing effort) or surveys (Deriso et al., 1985; Maunder and Punt, 2013; Schaub et al., 2024). Our example will use JAGS to fit a relatively simple age-structured stock assessment model, but we note that most current assessments are done using flexible and powerful software such as AD Model Builder or Stock Synthesis (Fournier et al., 2012; Methot and Wetzell, 2013).

We assume a single fishery with gear selectivity that is constant over time. Our auxiliary data are survey catches by age and year, with a survey selectivity pattern that is also constant over time. The default selectivity pattern for the survey provides information mostly on younger ages, whereas the fishery primarily exploits older fish. Thus a combined analysis is able to make better inferences than either data set alone. Essentially the goal of the analysis is to find the population estimates that are most consistent with the observed catches and auxiliary data.

The simulation code builds on that used in the cohort-specific catch-curve analysis (Section 6.1.2) and recruitment (Chapter 9). The number of ages is arbitrary but should be sufficient to characterize the selectivity patterns for the fishery and survey. Annual instantaneous fishing mortality rate varies randomly between user-set bounds. The instantaneous natural mortality rate can be difficult to estimate internally (Quinn and Deriso, 1999; Schaub et al., 2024), so here it is fixed in the simulation and the fitted model (e.g., using external information such as longevity: Lorenzen, 1996; Then et al., 2015).

```
rm(list=ls()) # Clear Environment

# Simulation to create survey and fishery catch-at-age matrices

A <- 8 # Arbitrary number of ages and years (but enough to fit model)
Y <- 12

LowF <- 0.1 # Arbitrary lower bound
HighF <- 0.5 # Arbitrary upper bound

AnnualF <- runif(Y, min=LowF, max=HighF)
AnnualM <- 0.2 # Fixed

# F by age, using separable model
Fishery_k <- 2 # Slope for logistic function
Fishery_a_50 <- 4 # Age at 0.5 selectivity
SelF <- array(data=NA, dim=A) # Fishery selectivity pattern
F_a <- array(data=NA, dim=c(A,Y)) # Matrix for F by age and year
Z_a <- array(data=NA, dim=c(A,Y)) # Matrix for Z by age and year
Exp_C_f <- array(data=NA, dim=c(A,Y)) # Fishery catch-at-age matrix
C_f <- array(data=NA, dim=c(A,Y)) # Obs fishery catch-at-age matrix
Exp_C_s <- array(data=NA, dim=c(A,Y)) # Survey catch-at-age matrix
C_s <- array(data=NA, dim=c(A,Y)) # Obs survey catch-at-age matrix
S_a <- array(data=NA, dim=c(A,(Y-1))) # Matrix for survival rate
for (a in 1:A){
  SelF[a] <- 1/(1+exp(-Fishery_k*(a-Fishery_a_50)))
} #a
#plot(SelF)
for (y in 1:(Y)) {
  for (a in 1:A){
    F_a[a,y] <- AnnualF[y] * SelF[a]
  } #a
} #y
Z_a <- F_a + AnnualM
S_a <- exp(-Z_a)

MeanR <- 1E5 # Arbitrary mean initial cohort size
N <- array(data=NA, dim=c(A, Y))
V_rec <- 0.1 # Log-scale recruitment variance
SD_rec <- sqrt(V_rec)
N[1,] <- trunc(MeanR * rlnorm(n=Y, 0, SD_rec))
```

```
# Starting size of each cohort at first age
for (a in 2:A){
  N[a,1] <- trunc(prod(S_a[1:(a-1),1])*MeanR * rlnorm(n=1, 0, SD_rec))
  # Use year-1 S_a to generate year-1 N
  for (y in 2:Y){
    N[a,y] <- rbinom(1, N[(a-1), (y-1)], S_a[(a-1),(y-1)])
  } #y
} #a


for (y in 1:(Y-1)){ # Deaths due to fishing (binomial random variate)
  for (a in 1:(A-1)){
    Exp_C_f[a,y] <- rbinom(1, N[a,y]-N[(a+1),(y+1)],(F_a[a,y]/Z_a[a,y]))
  } #y
} #a
for (y in 1:(Y-1)){ # Random catch last age
  Exp_C_f[A,y] <- rbinom(1,N[A,y],F_a[A,y]*(1-exp(-Z_a[A,y]))/Z_a[A,y])
} #y
for (a in 1:A){ # Random catch last year
  Exp_C_f[a,Y] <- rbinom(1,N[a,Y],F_a[a,Y]*(1-exp(-Z_a[a,Y]))/Z_a[a,Y])
} #y
for (y in 1:Y){
  C_f[,y] <- as.vector(t(rmultinom(n=1, size=sum(Exp_C_f[,y]),
                                   prob=Exp_C_f[,y])))
} #y


# Generate survey catch-at-age matrix
Survey_q <- 1E-3 # Catchability coefficient (Adjust relative to MeanR)
Survey_k <- -2 # Slope for logistic function
Survey_a_50 <- 4 # Age at 0.5 selectivity
SelS <- array(data=NA, dim=A) # Survey selectivity pattern
for (a in 1:A){
  SelS[a] <- 1/(1+exp(-Survey_k*(a-Survey_a_50)))
} #a
#plot(SelS)


for (y in 1:Y) {
  for (a in 1:A){
    Exp_C_s[a,y] <- Survey_q*SelS[a]*N[a,y]
  } #a
} #y


for (y in 1:Y){
  C_s[,y] <- as.vector(t(rmultinom(n=1, size=sum(Exp_C_s[,y]),
                                   prob=Exp_C_s[,y])))
} #y
```

Survival across ages and years varies randomly according to a binomial distribution (Section 3.1.2). Age- and year-specific catch also varies randomly according to a binomial distribution, based on the fraction of deaths due to fishing (F/Z). Sampling variation in the fishery and survey catch matrices is based on a multinomial distribution (Section 3.1.3). Compare the fishery and survey catch matrices to see the effects of gear selectivity. Also examine the population matrix to see strong and weak year classes moving through the population.

```
# Load necessary library packages
library(rjags)
library(R2jags)

# JAGS code ##################################
sink("CatchAge.txt")
cat("
model {
# Model parameters: AnnualF[1:Y], k and a_50 for fishery and survey,
# survey q, Mean recruitment, SD, and annual deviations years 1:Y,
# initial numbers ages 2:A for year 1

 Mean_lnR.est ~ dnorm(0, 1E-6)
   # Recruitment uses ln-scale parameter
 MeanR.est <- exp(Mean_lnR.est)
 V_rec.est ~ dunif(0, 10)
 tau <- 1/V_rec.est
 for(y in 1:Y) {
    Recruit.est[y] ~ dlnorm(Mean_lnR.est,tau)
     # Hierarchical structure for recruitment
    N.est[1,y]<-Recruit.est[y] }

    for(a in 2:A){
      Initial.est[a-1] ~ dlnorm(0, 1E-6)
        # Offset index to run Initials vector from 1 to A-1
    N.est[a,1]<-Initial.est[a-1] }

Fishery_k.est ~ dnorm(0, 1E-3)
  # Allow for positive or negative slope for logistic function
Fishery_a_50.est ~ dunif(0,A)
for (a in 1:A){
  SelF.est[a] <- 1/(1+exp(-Fishery_k.est*(a-Fishery_a_50.est)))
  }#a
```

```
for (y in 1:Y){
  AnnualF.est[y] ~ dunif(0,2) # Arbitrary range
  for (a in 1:A){
    F_a.est[a,y] <- AnnualF.est[y] * SelF.est[a]
      # Separable model (F[a,y]=Sel[a]*F[y])
    Z_a.est[a,y] <- F_a.est[a,y] + AnnualM
      # Assume that M is known (fixed externally)
  }#a
}#y

Survey_q.est ~ dgamma(1,1) # Uninformative non-negative prior
Survey_k.est ~ dnorm(0, 1E-3)
  # Allow for positive or negative slope for logistic function
Survey_a_50.est ~ dunif(0, A)
for (a in 1:A){
  SelS.est[a] <- 1/(1+exp(-Survey_k.est*(a-Survey_a_50.est)))
  } #a

  # Population projection
  for(a in 2:A) {
  for(y in 2:Y) {
    N.est[a,y] ~ dbin(exp(-AnnualM-F_a.est[(a-1),(y-1)]),
                    trunc(N.est[a-1,y-1])) } }

  #Likelihoods
  for (a in 1:A) {
  for (y in 1:Y) {
    C_f.est[a,y] <- N.est[a,y]*F_a.est[a,y]*(1-exp(-Z_a.est[a,y]))
                    /Z_a.est[a,y]
    C_f[a,y] ~ dpois(C_f.est[a,y])
    C_s.est[a,y] <- N.est[a,y]*Survey_q.est*SelS.est[a]
    C_s[a,y] ~ dpois(C_s.est[a,y])
  } } # y, a
  for (y in 1:Y){
    N.est.total[y] <- sum(N.est[,y])
  }

}
    ",fill=TRUE)
sink()

# Bundle data
jags.data <- list("Y", "A", "C_f", "C_s", "AnnualM")
```

```r
# Initial values
jags.inits <- function(){ list(Mean_lnR.est=rlnorm(n=1, meanlog=10,
                                                    sdlog=1),
                               Recruit.est=rlnorm(n=Y, meanlog=10,
                                                  sdlog=1),
                               Initial.est=rlnorm(n=(A-1), meanlog=10,
                                                  sdlog=1),
                               Survey_q.est=rgamma(n=1, shape=1,
                                                   scale=1),
                               V_rec.est=runif(n=1, min=0, max=10),
                               AnnualF.est=runif(n=Y, min=0, max=2))}


model.file <- 'CatchAge.txt'

# Parameters monitored
jags.params <- c("AnnualF.est", "Recruit.est", "Initial.est",
                 "SelF.est", "Survey_q.est", "MeanR.est",
                 "V_rec.est", "SelS.est", "N.est.total")

# Call JAGS from R
jagsfit <- jags(data=jags.data, jags.params, inits=jags.inits,
                n.chains = 3, n.thin=10, n.iter = 200000,
                model.file)
print(jagsfit)
plot(jagsfit)

par(mfrow = c(1, 2))
plot(seq(1:Y), jagsfit$BUGSoutput$median$AnnualF.est, ylab="Annual F",
     xlab="Year", type="b", pch=19,
     ylim=range(AnnualF,jagsfit$BUGSoutput$median$AnnualF.est))
points(AnnualF, lty=3, type="b")

N.total <- array(data=NA, dim=Y)
for (y in 1:Y){
  N.total[y] <- sum(N[,y])
}
plot(seq(1:Y), jagsfit$BUGSoutput$median$N.est.total, ylab="Total N",
     xlab="Year", type="b", pch=19,
     ylim=range(N.total,jagsfit$BUGSoutput$median$N.est.total))

points(N.total[], lty=3, type="b")

plot(seq(1:A), jagsfit$BUGSoutput$median$SelF.est,
     ylab="Fishery selectivity", xlab="Age", type="b", pch=19,
```

```
      ylim=c(0,1))
points(SelF, lty=3, type="b")

plot(seq(1:A), jagsfit$BUGSoutput$median$SelS.est,
     ylab="Survey selectivity", xlab="Age", type="b", pch=19,
     ylim=c(0,1))
points(SelS, lty=3, type="b")
```

Convergence (Rhat < 1.1) for this model requires a very large number of updates (and considerable time). Thinning is used to reduce the storage required. The plots following the JAGS code show that there is typically close agreement between the true and estimated values (Figure 10.2). It is essential to keep in mind that this analysis is being done with an assumed rate of natural mortality. The results of such an analysis, including credible intervals, are conditional on the assumed value. In this case, natural mortality is fixed at the correct value, so these results show the best case in terms of accuracy and precision.

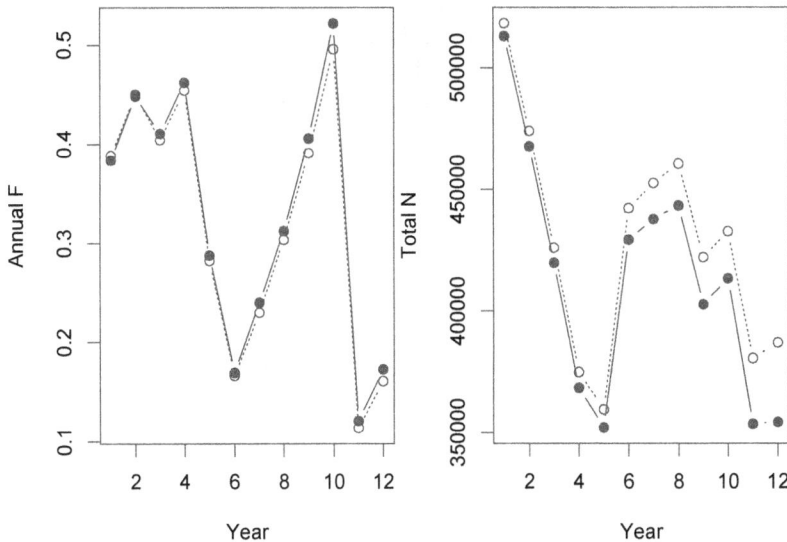

FIGURE 10.2 Estimates of fishing mortality (left) and total abundance (right) from a stock assessment model fitted to fishery and survey catch-at-age matrices. True values and model estimates are shown as dotted and solid lines, respectively.

Recruitment in this simulation was assumed to vary about a fixed mean, but the model could be extended to include a relationship between spawning stock and recruitment, as in Section 9.1. The analytical approach could then be modified to estimate the spawning stock-recruitment parameters internally (Brooks and Deroba, 2015).

When experimenting with alternative simulation settings, approximate estimates can be obtained using fewer updates (e.g., 20,000). Rhat values will typically exceed 1.1, but the estimates should show reasonable agreement with the true values. Final estimates can then be obtained using a sufficiently large value for the number of updates.

10.3   Exercises

1. For the two-age model (Section 10.1), compare true and estimated total abundance using absolute relative errors (abs((y-y.hat)/y)) for five trials using capture probabilities of 0.1, 0.2, and 0.3. Would you recommend a field trial using a capture probability of 0.1?

2. For the stock assessment model (Section 10.2), set the number of updates at a modest level (e.g., 20,000) and compare the level of convergence when fishing mortality is consistently low (e.g., 0.0-0.1) or high (1.0-1.1). Why is there a difference in how readily convergence is achieved for the two cases?

11

Final Thoughts

If you made it this far, a few themes should be apparent. The first is the important role of simulation in planning and conducting field research. Even writing the code to simulate a field study forces one to think about the design and parameters such as sample size that are under the investigator's control. Once the combined code for simulation and analysis is working, runs under different settings can show what is required to obtain useful results, at least for the simulated (ideal) case. The simulations and models used in this book can also be extended to include additional sources of error, such as ageing error or non-reporting or shedding of tags. This planning approach will also require some thought about what *is* a useful result, for example, a tolerable credible interval width or relative error $\left(\frac{\hat{\delta}-\delta}{\delta}\right)$. Upfront simulation and planning improves the odds of a successful field study, or perhaps prevents a waste of time and money conducting a study that would not address management needs.

Another theme has been the value of JAGS in constructing and fitting fisheries models. In my experience, BUGS-family code is a valuable teaching tool, helping students visualize the models and how parameter estimates are obtained. Writing and running JAGS code is vastly superior to a black-box analysis from a learning perspective. From a research perspective, JAGS code can be tailored to the specific characteristics of a field study. Examples in this book hopefully have illustrated how auxiliary studies can be folded into a Bayesian analysis to improve parameter estimates. JAGS software is still popular and being actively developed, but there are now a number of options for Bayesian analysis (Štrumbelj et al., 2024). A new book by Kéry and Kellner (2024) on Bayesian model fitting using JAGS, Nimble, and Stan should be a good resource for exploring the strengths and weaknesses of those software packages. In any case, experience with JAGS should provide a strong foundation for future fisheries modeling, even if Bayesian analyses transition to next-generation software.

A third theme is the importance of learning by doing. Ideally, this includes not only writing code for simulation and analysis, but also carrying out the field work that builds experience and ecological intuition. Within the scope of this book, learning by doing means running and modifying code in order to understand fully how each analysis works. JAGS and especially R code can sometimes be cryptic, so I have encouraged running sections of code or even parts of a line of code in order to gain understanding. One of the best things about recent books on Bayesian analysis is the routine inclusion of code. Seeing, running and modifying the code is a great help in understanding how these models work.

This is intended to be an introductory text, so some important topics were not covered or only mentioned in passing. The binomial-mixture model (Section 5.3) is a simple example of a hierarchical model, but the approach is much broader. Hierarchical modeling can be used to separate the observation process from the underlying ecological process (latent state) that is of primary interest (Royle and Dorazio, 2008). Kéry and Schaub (2012) illustrate this approach in a hierarchical population model for ibex (*Capra ibex*). Annual counts include both process error and observation error, so a model that separates the two error sources provides more biologically relevant estimates of population parameters (Kéry and Schaub, 2012). Recent publications using Bayesian hierarchical models have often been terrestrial studies (e.g., Johnson et al., 2010), but the concepts and analytical methods are entirely transferable to fisheries systems.

I have also not included some traditional fisheries models, such as the biomass dynamic or surplus production models that have been widely used for fish stock assessment (Hilborn and Walters, 1992). These models have been popular because they are simple, require limited data, and provide an estimate of maximum sustainable yield. Early versions often assumed equilibrium conditions that were rarely met (Hilborn and Walters, 1992). Recent versions using non-equilibrium models (Meyer and Millar, 1999; Winker et al., 2018) illustrate the potential of a Bayesian analysis, including a state-space formulation for sorting out observation and process error and use of informative priors.

Another general fisheries topic not covered is yield or harvest modeling. We have focused on estimating population parameters, so the logical next step would be to use those parameter estimates to explore management options. For example, a yield model might be used to explore the potential benefits of regulation. The yield model would use information about fishing and natural mortality (Chapter 7), growth (Chapter 8), and possibly recruitment (Chapter 9). The yield model could be constructed within an IPM (Chapter 10), so that the uncertainty associated with the various population parameters would not be lost. For a commercial fishery, the yield model would typically be in terms of weight, corresponding to how income is derived. The model could be used to examine the potential gains and losses from management options such as a quota or seasonal closure. For a recreational fishery, a yield model might focus on numbers (catch rate) rather than weight, or on the proportion of fish within "memorable" and "trophy" size ranges. Management for a recreational fishery is more likely to use creel and size regulations than quotas and could, for example, compare the benefits of various minimum and slot size limits. A recent article by Doll et al. (2021) perfectly illustrates these concepts using a Bayesian IPM fitted with JAGS.

Kéry (2010) closed his book with a prediction that, for many readers, WinBUGS will have "changed forever the way in which you think about, and conduct, statistical modeling." WinBUGS certainly started that process for me, followed by other BUGS-family variants including JAGS. It will be interesting to see the future contributions of Bayesian software packages to fisheries science and management.

A

Empirical Examples

The purpose of this appendix is to show, for a subset of models, the process of model fitting to real values rather than simulated data. Removing the simulation code generally shortens and simplifies the examples. I have included brief comments regarding necessary code modifications. Reported point estimates are medians, which provide a better measure of central tendency when posterior distributions are skewed (Kruschke, 2021).

A.1 Abundance: Removal

The removal method (Section 5.1) is used to estimate abundance in a closed study area, based on a sequence of catches. Otis et al. (1978) provided an example for whitefish (*Coregonus clupeaformis*) in Shakespeare Island Lake, based on gill netting for 7 successive weeks (original data from Ricker, 1958). No changes were needed in the R or JAGS code other than replacing the simulation code with the vector of observed catches and changing the number of removal samples.

```r
rm(list=ls()) # Clear Environment

# Removal study - whitefish (Ricker 1958)
Catch <- c(25, 26, 15, 13, 12, 13, 5)
N.removals <- length(Catch) # Number of removals

# Load necessary library packages
library(rjags)    # Package for fitting JAGS models from within R
library(R2jags)   # Package for fitting JAGS models. Requires rjags

# JAGS code for fitting model
sink("RemovalModel.txt")
cat("
```

```
model{

# Priors
 CapProb.est ~ dunif(0,1)
 N.est ~ dunif(TotalCatch, 2000)

 N.remaining[1] <- trunc(N.est)
 for(j in 1:N.removals){
     Catch[j]~dbin(CapProb.est, N.remaining[j]) # jth removal
     N.remaining[j+1] <- N.remaining[j]-Catch[j]
  } #j

}
     ",fill=TRUE)
sink()

# Bundle data
TotalCatch <- sum(Catch[])
jags.data <- list("Catch", "N.removals", "TotalCatch")


# Initial values
jags.inits <- function(){ list(CapProb.est=runif(1, min=0, max=1),
                               N.est=runif(n=1, min=TotalCatch,
                                            max=2000))}


model.file <- 'RemovalModel.txt'

# Parameters monitored
jags.params <- c("N.est", "CapProb.est")

# Call JAGS from R
jagsfit <- jags(data=jags.data, jags.params, inits=jags.inits,
                n.chains = 3, n.thin = 1, n.iter = 40000,
                model.file)
print(jagsfit)
plot(jagsfit)
```

The analysis by Otis et al. (1978) using program CAPTURE indicated that a constant capture probability model was appropriate. Their estimate ($\hat{N}$=138.07) was similar to the median estimate I obtained using Bayesian methods (144.4). Their symmetrical 95% confidence interval (109-167) was narrower than the 95% credible interval I obtained using Bayesian methods

(121.8-225.3), due to the right-skew of the posterior distribution for N.est (e.g., try `hist(jagsfit$BUGSoutput$sims.list$N.est)`).

A.2   Abundance: Mark-recapture

Ogle (2016) analyzed a two-sample mark-recapture experiment for walleye larger than 304 mm (*Sander vitreus*). No changes in the R or JAGS code were required other than replacing the Section 5.2 simulation code with observed values.

```
rm(list=ls()) # Clear Environment

# Ogle (2016)
# Wisconsin Department of Natural Resources
# Mark-recapture walleye > 304 mm
n1 <- 2555 # First sample (Marked)
n2 <- 274 # Second sample
m2 <- 92 # Marked fish in second sample

# Load necessary library packages
library(rjags)    # Package for fitting JAGS models from within R
library(R2jags)   # Package for fitting JAGS models. Requires rjags

# JAGS code
sink("TwoSampleCR.txt")
cat("
    model {

    # Priors
    MarkedFraction ~ dunif(0, 1)

    # Calculated value
    N.hat <- n1 /MarkedFraction

    # Likelihood
    # Binomial distribution for observed recaptures
    m2 ~ dbin(MarkedFraction, n2)
    }

    ",fill = TRUE)
```

```
sink()

# Bundle data
jags.data <- list("n1", "n2", "m2")

# Initial values.
jags.inits <- function(){ list(MarkedFraction=runif(1, min=0, max=1))}

model.file <- 'TwoSampleCR.txt'

# Parameters monitored
jags.params <- c("N.hat", "MarkedFraction")

# Call JAGS from R
jagsfit <- jags(data=jags.data, inits=jags.inits, jags.params,
                n.chains = 3, n.thin = 1, n.iter = 2000,
                n.burnin = 1000, model.file)
print(jagsfit, digits=3)
plot(jagsfit)
```

Results using Bayesian methods (N.est median=7569.6, 95% credible interval 6515.7-9015.2) were similar to estimates provided by Ogle (2016) using the Chapman (1951) estimator ($\hat{N}$=7557, 95% confidence interval 6454-8973).

A.3   Survival: Age-composition

Section 6.1 showed an approach for estimating survival using age-composition data. Here I illustrate a different approach: a Bayesian version of a traditional catch curve analysis. Quinn and Deriso (1999) provided a catch curve analysis for Pacific halibut in the northeastern Pacific Ocean, using fishery catch data for the 1960 year class between ages 10 and 20 (i.e., tracking a single cohort across multiple years).

Unlike the approach in Section 6.1, a traditional catch curve analysis uses ln-transformed catches, for the following reason. The equation for population size at time t (Section 7.2) is $N_t = N_0 * exp(-Z*t)$. Multiplying both sides by a time-independent exploitation rate μ produces $\mu * N_t = \mu * N_0 * exp(-Z*t)$ or $C_t = \mu * N_0 * exp(-Z*t)$ (Quinn and Deriso, 1999). Log-transforming that equation produces $ln(C_t) = ln(\mu * N_0) - Z*t$, which is a linear regression equation with intercept $ln(\mu * N_0)$ and slope -Z. Fitting the model to log-transformed catches implies that the catch data are considered to be lognormally distributed (Quinn and Deriso, 1999).

The Bayesian code uses coded ages 1-10 rather than true ages 10-20 and fits the linear regression model using the dnorm() function for the likelihood. Estimated survival is obtained as $exp(-Z)$, and as always the MCMC process provides a complete characterization of that calculated parameter.

```
rm(list=ls()) # Clear Environment

# Estimating survival using fishery catch data
# Pacific halibut, ages 10-20 (1960 year class)
# Quinn and Deriso 1999

Catch <- c(179187, 122785, 89241, 58111, 26682, 23339,
            14668, 10050, 5866, 3121, 4837)
lnC <- log(Catch) # ln-transform to fit traditional catch curve
Age <- 10:20 # True ages for plotting
MaxCode <- length(Catch)
CodedAge <- 1:MaxCode # Coded ages for fitting model

#---------------------------------------------------
# Fit ln-scale model
# Load necessary library packages
library(rjags)
library(R2jags)

# JAGS code
sink("CatchCurve.txt")
cat("
model{
    # Priors
    Z.est ~ dunif(0, 2) # Instantaneous total mortality rate
    lnC.intercept ~ dunif(0, 20)
    Var.est ~ dunif(0, 20)
    tau <- 1/Var.est # Precision
    sd.est <- sqrt(Var.est)
    S.est <- exp(-Z.est)

    # Likelihood
    for (a in 1:MaxCode){
      lnC.hat[a] <- lnC.intercept - (Z.est*a)
      lnC[a] ~ dnorm(lnC.hat[a], tau)
    } # a
}
    ",fill=TRUE)
sink()
```

```
# Bundle data
jags.data <- list("lnC", "MaxCode")

# Initial values
jags.inits <- function(){ list(Z.est=runif(n=1, min=0, max=2),
                               lnC.intercept=runif(n=1, min=0, max=20),
                               Var.est=runif(n=1, min=0, max=20))}


model.file <- 'CatchCurve.txt'

# Parameters monitored
jags.params <- c("S.est", "Z.est", "lnC.intercept",
                 "sd.est", "lnC.hat")

# Call JAGS from R
jagsfit <- jags(data=jags.data, jags.params, inits=jags.inits,
                n.chains = 3, n.thin=1, n.iter = 4000, n.burnin=2000,
                model.file)
print(jagsfit)
plot(jagsfit)

y.bounds <- range(lnC, jagsfit$BUGSout$median$lnC.hat)
plot(Age, lnC, ylim=y.bounds, xlab="Age", ylab="ln-Catch")
points(Age, jagsfit$BUGSout$median$lnC.hat, type="l", col="red")
```

Quinn and Deriso (1999) reported an estimated Z of 0.409 with a 95% confidence interval of 0.359-0.459; our Bayesian median estimate was 0.412, with 95% credible interval of 0.353-0.469. Quinn and Deriso (1999) estimated survival to be 0.664, with 95% confidence interval of 0.632-0.698. The Bayesian median estimate was 0.663, with 95% credible interval of 0.626-0.703. This is yet another example where Bayesian results using uninformative priors are very similar to frequentist results. The last three lines of R code show the ln-transformed catches and fitted curve. The plot is essentially identical to that shown by Quinn and Deriso (1999). The ln-transformed catches are well described by the fitted curve, suggesting that the assumptions of a catch curve analysis are reasonable for the period of analysis.

A.4   Survival: Tag-based

Section 6.3 showed an approach for estimating survival rate using a multi-period tag-return study. Here I have replaced the simulation code with a male wood duck (*Aix sponsa*) band-recovery matrix [Brownie et al. (1985); table 2.2]. The band recovery values are entered as a vector, then transformed using an upper diagonal matrix using the matrix() function. Variables were renamed to be more consistent with the band-recovery experiment. The main change to the R code was that banded wood ducks were released for 3 years (1964-1966), but bands were recovered for 5 years (1964-1968). This required separate variables for the number of release (RelYears) and recovery years (RecYears), unlike the 3X3 example from Section 6.3. Brownie et al. (1985) fitted a model with time-dependent survival and tag-recovery rates, but here I estimated time-independent survival and tag-recovery rates for consistency with Section 6.3. The only changes required for the JAGS code were because of the different number of years for release and recovery.

```r
rm(list=ls()) # Clear Environment

# Recovered bands of adult male wood ducks
# Brownie et al. (1985)

RelYears <- 3 # 1964-1966 # Release years
RecYears <- 5 # 1964-1968 # Recovery years
R <- c(1603, 1595, 1157) # Releases by year
Recoveries <- c(127, 44, 37, 40, 17, NA, NA, 62, 76,
                44, 28, NA, NA, NA, 82, 61, 24, NA)
# Convert vector of band recoveries into upper diagonal
# matrix, NA added for final column for bands not seen again
Rec.Mat <- matrix(Recoveries,nrow = RelYears,ncol = (RecYears+1),
                byrow=TRUE)
# Placeholder NA - replace by number of bands not seen again
for (y in 1:RelYears){
  Rec.Mat[y,RecYears+1] <- R[y]-sum(Rec.Mat[y,y:RecYears])
} #y

# Load necessary library packages
library(rjags)
library(R2jags)

# Specify model in JAGS
sink("TagReturn.txt")
cat("
model {
  # Priors
  S.est ~ dunif(0,1) # Time-independent survival rate
  r.est ~ dunif(0,1) # Time-independent tag-recovery rate
```

```
  # Calculated value
  A.est <- 1-S.est # Rate of mortality

# Cell probabilities
for (i in 1:RelYears) {
   p.est[i,i] <- r.est
   for (j in (i+1):RecYears) {
      p.est[i,j] <- p.est[i,(j-1)] * S.est
      } #j
   p.est[i,RecYears+1] <- 1 - sum(p.est[i, i:RecYears])
   # Last column is prob of not being seen again
   } #i

# Likelihood
  for (i in 1:RelYears) {
    Rec.Mat[i,i:(RecYears+1)] ~ dmulti(p.est[i,i:(RecYears+1)], R[i])
    }#i
 } # model

",fill=TRUE)
sink()

# Bundle data
jags.data <- list("Rec.Mat", "RelYears", "RecYears", "R")

S.init <- runif(1,min=0,max=1)
r.init <- 1-S.init  # Initialize r relative to S

# Initial values
jags.inits <- function(){list(S.est = S.init, r.est=r.init)}

model.file <- 'TagReturn.txt'

# Parameters monitored
jags.params <- c("S.est", "r.est")

# Call JAGS from R
jagsfit <- jags(data=jags.data, jags.params, inits=jags.inits,
                n.chains = 3, n.thin=1, n.iter = 4000, n.burnin=2000,
                model.file)
print(jagsfit)
plot(jagsfit)
```

The estimates from the Bayesian analysis (median survival 0.66, 95% credible interval 0.62-0.71; median tag-recovery rate 0.06, 95% credible interval 0.06-0.07) were very similar to maximum-likelihood estimates (mean survival 0.64, 95% confidence interval 0.56-0.71; mean tag-recovery rate 0.06, 95% confidence interval 0.05-0.07) reported by Brownie et al. (1985).

A.5 Growth: age:length

This example data set for fitting a von Bertalanffy growth curve is for rougheye rockfish (*Sebastes aleutianus*). The original study was by Nelson and Quinn II (1987); Quinn and Deriso (1999) fitted the model using nonlinear least squares.

Using real rather than simulated data usually results in simpler code, but here, the values provided by Quinn and Deriso (1999) were age-specific sample sizes, averages, and standard deviations. A curve could be fitted to age-specific averages, but that gives equal weight to each age, whereas age-specific sample sizes actually varied widely (range from 1 to 31). To approximate the original data set of individual fish age and length observations, I looped over the sample size by age and generated "observed" values using the rnorm() function. Working with individual observations retains the variation in the original length data set. Note that the "observed" data set will vary from run to run, so it is helpful to run the code multiple times. The JAGS code was unchanged from the example in Section 8.1.

```
# Fitting von Bertalanffy curve
# Rougheye rockfish
# Nelson, B., and T. J. Quinn II. 1987. Population parameters for
# rougheye rockfish (Sebastes aleutianus). Proc. Int. Rockfish
# Symp., Alaska Sea Grant Rep. 87-2:209-228.
rm(list=ls()) # Clear Environment

AgeClass <- c(2:38, 40:47, 49, 50, 52:68, 70, 72,75, 77, 80, 82, 85, 95)
ave.Len <- c(18, 14.5, 17.9, 20.6, 21.9, 23.4, 25.2, 27.2, 29.6, 30.7,
            32.4, 33, 32.8, 35, 36.3, 37.2, 36.2, 38.8, 37.7, 40.6,
            40.9, 38.9, 40, 41.9, 42.7, 42.6, 45.5, 40.9, 44.6,
            46.6, 44.9, 47.2, 48.3, 45.4, 48.8, 46.5, 46.5, 47.9,
            49.3, 48, 49, 50, 51.3, 52, 48.9, 48.7, 48.6, 51.5, 52,
            50, 52.7, 59, 52, 50, 50, 51, 50, 52, 50, 59, 55, 52.5, 50,
            54, 59.5, 51, 51, 60, 74, 49, 57, 60.5)
sd.Len <- c(NA, 4.9, 2.2, 2.1, 2.1, 2.3, 2.5, 2, 3, 2.9, 3.7, 2.3,
```

```
           2.9, 4.1, 2.4, 5.1, 4.4, 1.8, 5.4, 3.1, 4.3, 4.6, 2.4,
           3.9, 5.5, 6.4, 4, 4.6, 4.4, 3.6, 6.3, 4.6, 5.1, 4.5,
           3.4, 4, 2.5, 4.9, 2.5, 2.1, 2.0, 2.2, 5.9, NA, 2.7,
           2.5, 2.8, 0.7, 11.4, NA, 8, NA, 3.9, NA, NA, 1.4, NA,
           1, NA, 9.9, 1.4, 2.1, 0, 1, 13.4, 1.4, NA, 12.7, NA,
           NA, NA, 16.3)
n.AgeClass <- c(1,6,19,31,27,20,11,19,12,13,12,4,8,11,7,12,10,5,9,7,14,
            14,13,15,15,9,8,8,13,11,9,9,16,13,10,14,4,8,3,7,3,7,10,
            1,9,3,7,2,4,1,3,1,5,1,1,4,1,3,1,2,2,2,2,1,2,2,1,2,1,1,
            1,2)
N.AgeLen <- sum(n.AgeClass[])
Len <- array(data=NA, dim=N.AgeLen)
Age <- array(data=NA, dim=N.AgeLen)
fish <- 1 # Counter for individual fish
for (a in 1:length(AgeClass)){
  if (n.AgeClass[a]==1) {
    Age[fish] <- AgeClass[a]
    Len[fish] <- ave.Len[a]
    fish <- fish+1
  } else
    for (j in 1:n.AgeClass[a]){
      Age[fish] <- AgeClass[a]
      Len[fish] <- rnorm(n=1, mean=ave.Len[a], sd=sd.Len[a])
      fish <- fish+1
    } #j
} #a

# Load necessary library packages
library(rjags)
library(R2jags)

# JAGS code
sink("GrowthCurve.txt")
cat("
model{

# Priors

 L_inf.est ~ dunif(0, 200)
 k.est ~ dunif(0, 2)
 t0.est ~ dunif(-5, 5)
 Var_L.est ~ dunif(0,100)    # Variability in length at age

# Calculated value
 tau.est <- 1/Var_L.est

# Likelihood
 for (i in 1:N.AgeLen) {
```

```
    Len_hat[i] <- L_inf.est*(1-exp(-k.est*(Age[i]-t0.est)))
    Len[i] ~ dnorm(Len_hat[i], tau.est)
  } #i
}
    ",fill=TRUE)
sink()

# Bundle data
jags.data <- list("N.AgeLen", "Age", "Len")

# Initial values
jags.inits <- function(){list(L_inf.est=runif(n=1, min=0, max=max(Len)),
                              k.est=runif(n=1, min=0, max=2),
                              t0.est=runif(n=1, min=-5, max=5),
                              Var_L.est=runif(n=1, min=0, max=100))}

model.file <- 'GrowthCurve.txt'

# Parameters monitored
jags.params <- c("L_inf.est", "k.est", "t0.est", "Var_L.est")

# Call JAGS from R
jagsfit <- jags(data=jags.data, jags.params, inits=jags.inits,
                n.chains = 3, n.thin=1, n.iter = 10000,
                model.file)
print(jagsfit)
plot(jagsfit)

Len_hat <- array(data=NA, dim=N.AgeLen)
for (a in 1:N.AgeLen){
  Len_hat[a] <- jagsfit$BUGSoutput$median$L_inf.est*(1-
                  exp(-jagsfit$BUGSout$median$k.est
                  *(Age[a]-jagsfit$BUGSout$median$t0.est)))
  } #a
plot(Age, Len, xlab="Age", ylab="Length (cm)")
points(Age, Len_hat, type="l", col="red")
```

Bayesian median estimates from one run (L_∞=53.761, k=0.052, t_0=−4.470) were in reasonable agreement with the least-squares estimates reported by Quinn and Deriso (1999): L_∞=54.9, k=0.0441, and t_0=−4.53. The estimated variance of points about the fitted line (Var_L.est=17.910) was similar to the residual mean square of 17.13 reported by Quinn and Deriso (1999), suggesting that the simulation approach used here was effective in restoring the variability in individual age:length observations.

A.6   Recruitment: Beverton-Holt curve

Section 9.1 used a complex age-structured simulation model to produce spawning stock and recruitment values. It is much simpler to use a real data set, as illustrated here for North Sea plaice (Ricker, 1975). Spawning stock and recruitment values are paired in Ricker (1975), so no time lag between indices is needed. The only changes from the R and JAGS code used in Section 9.1 were to replace the simulation code with fixed values, remove the time lag in indices, and to use Ricker's notation for spawning stock (P) and recruitment (R).

```
# Fitting Beverton-Holt stock-recruitment curve
# North Sea plaice. Ricker 1975, Table 11.8
# Units arbitrary

# Adult stock biomass (P)
P <- c(9, 9, 9, 10, 10, 10, 10, 10, 11, 11, 12, 12, 18, 18, 19, 20,
       21, 21, 26, 32, 35, 45, 54, 70, 82, 88)
R <- c(13, 20, 45, 13, 13, 20, 21, 26, 11, 12, 8, 15, 7, 10, 17,
       16, 11, 15, 16, 33, 10, 23, 13, 13, 12, 24)

# Load necessary library packages
library(rjags)
library(R2jags)

sink("StockRecruit.txt")
cat("
model {
# Model parameters: Beverton-Holt alpha and beta,
# ln-scale variance for fitted curve

# Priors
 BH_alpha.est ~ dunif(0, 10)
 BH_beta.est ~ dunif(0, 10)
 V_rec.est ~ dunif(0, 10)
 tau <- 1/V_rec.est

# Likelihood
    for(y in 1:Y) {
      lnR_hat[y] <- log(1/(BH_alpha.est+BH_beta.est/P[y]))
      R[y] ~ dlnorm(lnR_hat[y],tau)
```

```
        R_hat[y] <- exp(lnR_hat[y])
        } #y
}
    ",fill=TRUE)
sink()


# Bundle data
Y <- length(P)
jags.data <- list("Y", "P", "R")

# Initial values
y <- R[1:Y]
x <- P[1:Y]
nls_init <- nls(y ~ 1/(BH_a_start+BH_b_start/x),
                start = list(BH_a_start = 0.001, BH_b_start = 0.001),
                algorithm="port",
                lower=c(1E-6, 0), upper=c(1000,1000))
#summary(nls_init)
nls_save <- coef(nls_init)
#y_hat <- 1/(nls_save[1]+nls_save[2]/x)
#plot(x,y)
#points(x,y_hat, col="red", type="l")


jags.inits <- function(){ list(BH_alpha.est=nls_save[1],
                               BH_beta.est=nls_save[2],
                               V_rec.est=runif(n=1, min=0, max=10))}

model.file <- 'StockRecruit.txt'

# Parameters monitored
jags.params <- c("BH_alpha.est", "BH_beta.est", "V_rec.est",
                 "R_hat")

# Call JAGS from R
jagsfit <- jags(data=jags.data, jags.params, inits=jags.inits,
                n.chains = 3, n.thin=1, n.iter = 10000,
                model.file)
print(jagsfit)
plot(jagsfit)

df.xy <- data.frame(P[1:Y], R[1:Y],
                    jagsfit$BUGSoutput$median$R_hat)
df.xy <- df.xy[order(df.xy[,1]),]  # Sort by survey index for plotting
y.bounds <- range(R[1:Y],jagsfit$BUGSoutput$median$R_hat)
```

```
plot(df.xy[,1],df.xy[,2], ylab="Recruitment index",
    xlab="Spawning Stock Index", ylim=y.bounds) #c(min.y, max.y))
points(df.xy[,1], df.xy[,3], type="l")
```

Bayesian estimates of α (median=0.060, 95% credible interval 0.042-0.075) and β (0.091, 95% credible interval 0.003-0.368) were similar to estimates (0.06312; 0.1606) provided by Ricker (1975).

Bibliography

Allen, M. S. (1997). Effects of variable recruitment on catch-curve analysis for crappie populations. *North American Journal of Fisheries Management*, 17(1):202–205.

Banner, K. M., Irvine, K. M., and Rodhouse, T. J. (2020). The use of Bayesian priors in Ecology: The good, the bad and the not great. *Methods in Ecology and Evolution*, 11(8):882–889.

Beverton, R. J. H. and Holt, S. J. (1993). *On the Dynamics of Exploited Fish Populations*. Springer Netherlands, Dordrecht.

Bolker, B. M. (2008). *Ecological Models and Data in R*. Princeton University Press, Princeton, New Jersey, 1st edition.

Brooks, E. N. and Deroba, J. J. (2015). When "data" are not data: The pitfalls of post hoc analyses that use stock assessment model output. *Canadian Journal of Fisheries and Aquatic Sciences*, 72(4):634–641.

Brooks, S. P. and Gelman, A. (1998). General methods for monitoring convergence of iterative simulations. *Journal of Computational and Graphical Statistics*, 7(4):434–455.

Brownie, C., Anderson, D., Burnham, K., and Robson, D. (1985). Statistical inference from band recovery data: A handbook. Federal Government Series 131, U.S. Fish and Wildlife, Washington, D.C.

Bryant, M. D. (2000). Estimating fish populations by removal methods with minnow traps in southeast Alaska streams. *North American Journal of Fisheries Management*, 20(4):923–930.

Burnham, K. P. and Anderson, D. R. (1998). *Model Selection and Inference*. Springer, New York, NY.

Campana, S. E. (2001). Accuracy, precision and quality control in age determination, including a review of the use and abuse of age validation methods. *Journal of Fish Biology*, 59(2):197–242.

Chapman, D. (1951). *Some Properties of the Hypergeometric Distribution with Applications to Zoological Sample Censuses*. University of California Press.

Conn, P. B., Diefenbach, D. R., Laake, J. L., Ternent, M. A., and White,

G. C. (2008). Bayesian analysis of wildlife age-at-harvest data. *Biometrics*, 64(4):1170–1177.

Conn, P. B., Johnson, D. S., Williams, P. J., Melin, S. R., and Hooten, M. B. (2018). A guide to Bayesian model checking for ecologists. *Ecological Monographs*, 88(4):526–542.

de Valpine, P., Turek, D., Paciorek, C. J., Anderson-Bergman, C., Lang, D. T., and Bodik, R. (2017). Programming with models: Writing statistical algorithms for general model structures with NIMBLE. *Journal of Computational and Graphical Statistics*, 26(2):403–413.

Deriso, R. B., Quinn II, T. J., and Neal, P. R. (1985). Catch-age analysis with auxiliary information. *Canadian Journal of Fisheries and Aquatic Sciences*, 42(4):815–824.

Doll, J. C. and Jacquemin, S. J. (2018). Introduction to Bayesian modeling and inference for fisheries scientists. *Fisheries*, 43(3):152–161.

Doll, J. C. and Jacquemin, S. J. (2019). Bayesian model selection in fisheries management and ecology. *Journal of Fish and Wildlife Management*, 10(2):691–707.

Doll, J. C., Wood, C. J., Goodfred, D. W., and Rash, J. M. (2021). Incorporating batch mark–recapture data into an integrated population model of brown trout. *North American Journal of Fisheries Management*, 41(5):1390–1407.

Dorazio, R. M. (2016). Bayesian data analysis in population ecology: Motivations, methods, and benefits. *Population Ecology*, 58(1):31–44.

Efron, B. and Tibshirani, R. J. (1993). *An Introduction to the Bootstrap*. Chapman & Hall/CRC, Boca Raton, Florida, 1st edition.

Ellison, A. M. (2004). Bayesian inference in ecology. *Ecology Letters*, 7(6):509–520.

Fielder, D. G. and Bence, J. R. (2014). Integration of auxiliary information in statistical catch-at-age (SCA) analysis of the Saginaw Bay stock of walleye in Lake Huron. *North American Journal of Fisheries Management*, 34(5):970–987.

Flowers, H. J. and Hightower, J. E. (2013). A novel approach to surveying sturgeon using side-scan sonar and occupancy modeling. *Marine and Coastal Fisheries: Dynamics, Management, and Ecosystem Science*, 5(1):211–223.

Flowers, H. J. and Hightower, J. E. (2015). Estimating sturgeon abundance in the Carolinas using side-scan sonar. *Marine and Coastal Fisheries: Dynamics, Management, and Ecosystem Science*, 7(1):1–9.

Fournier, D. A., Skaug, H. J., Ancheta, J., Ianelli, J., Magnusson, A., Maunder, M. N., Nielsen, A., and Sibert, J. (2012). AD Model Builder: Using automatic differentiation for statistical inference of highly parameterized complex nonlinear models. *Optimization Methods and Software*, 27(2):233–249.

Gelman, A. and Hill, J. (2007). *Data Analysis Using Regression and Multilevel/Hierarchical Models*. Cambridge University Press, Cambridge, 1st edition.

Gelman, A., Hwang, J., and Vehtari, A. (2014). Understanding predictive information criteria for Bayesian models. *Statistics and Computing*, 24(6):997–1016.

Gelman, A., Lee, D., and Guo, J. (2015). Stan: A probabilistic programming language for Bayesian inference and optimization. *Journal of Educational and Behavioral Statistics*, 40(5):530–543.

Gelman, A. and Rubin, D. B. (1992). Inference from iterative simulation using multiple sequences. *Statistical Science*, 7:457–472.

Hayes, D. B. and Brodziack, JKT. (1997). Reply: Efficiency and bias of estimators and sampling designs for determining lengthweight relationships of fish. *Canadian Journal of Fisheries and Aquatic Sciences*, 54(3):744–745.

Hearn, W. S., Hoenig, J. M., Pollock, K. H., and Hepworth, D. A. (2003). Tag reporting rate estimation: 3. Use of planted tags in one component of a multiple-component fishery. *North American Journal of Fisheries Management*, 23(1):66–77.

Hearn, W. S., Sandland, R. L., and Hampton, J. (1987). Robust estimation of the natural mortality rate in a completed tagging experiment with variable fishing intensity. *ICES Journal of Marine Science*, 43(2):107–117.

Hewitt, D. A., Janney, E. C., Hayes, B. S., and Shively, R. S. (2010). Improving inferences from fisheries capture-recapture studies through remote detection of PIT tags. *Fisheries*, 35(5):217–231.

Hightower, J. E. and Harris, J. E. (2017). Estimating fish mortality rates using telemetry and multistate models. *Fisheries*, 42(4):210–219.

Hightower, J. E., Loeffler, M., Post, W. C., and Peterson, D. L. (2015). Estimated survival of subadult and adult Atlantic sturgeon in four river basins in the southeastern United States. *Marine and Coastal Fisheries*, 7(1):514–522.

Hilborn, R. and Walters, C. J. (1992). *Quantitative Fisheries Stock Assessment*. Springer US, Boston, MA.

Hooten, M. B. and Hobbs, N. T. (2015). A guide to Bayesian model selection for ecologists. *Ecological Monographs*, 85(1):3–28.

Jiang, H., Pollock, K. H., Brownie, C., Hoenig, J. M., Latour, R. J., Wells, B. K., and Hightower, J. E. (2007). Tag return models allowing for harvest and catch and release: Evidence of environmental and management impacts on striped bass fishing and natural mortality rates. *North American Journal of Fisheries Management*, 27(2):387–396.

Johnson, H. E., Scott Mills, L., Wehausen, J. D., and Stephenson, T. R. (2010). Combining ground count, telemetry, and mark–resight data to infer population dynamics in an endangered species. *Journal of Applied Ecology*, 47(5):1083–1093.

Kéry, M. (2010). *Introduction to WinBUGS for Ecologists: Bayesian Approach to Regression, ANOVA, Mixed Models and Related Analyses*. Academic Press, Amsterdam, 1st edition.

Kéry, M. and Kellner, K. (2024). *Applied Statistical Modelling for Ecologists*. Elsevier, Amsterdam, 1st edition.

Kéry, M. and Schaub, M. (2012). *Bayesian Population Analysis Using Win-BUGS: A Hierarchical Perspective*. Academic Press, Amsterdam, 1st edition.

Kruschke, J. K. (2021). Bayesian Analysis Reporting Guidelines. *Nature Human Behaviour*, 5(10):1282–1291.

Law, A. M. and Kelton, W. D. (1982). *Simulation Modeling and Analysis*. McGraw-Hill, New York, NY.

Lemoine, N. P. (2019). Moving beyond noninformative priors: Why and how to choose weakly informative priors in Bayesian analyses. *Oikos*, 128(7):912–928.

Link, W. A. and Barker, R. J. (2010). *Bayesian Inference: With Ecological Applications*. Academic Press, London, 1st edition.

Link, W. A., Cam, E., Nichols, J. D., and Cooch, E. G. (2002). Of bugs and birds: Markov Chain Monte Carlo for hierarchical modeling in wildlife research. *The Journal of Wildlife Management*, 66(2):277–291.

Link, W. A. and Eaton, M. J. (2012). On thinning of chains in MCMC. *Methods in Ecology and Evolution*, 3(1):112–115.

Lorenzen, K. (1996). The relationship between body weight and natural mortality in juvenile and adult fish: A comparison of natural ecosystems and aquaculture. *Journal of Fish Biology*, 49(4):627–642.

Lunn, D., Jackson, C., Best, N., Thomas, A., and Spiegelhalter, D. (2012). *The BUGS Book: A Practical Introduction to Bayesian Analysis*. Chapman & Hall/CRC, Boca Raton, Florida, 1st edition.

Lunn, D., Spiegelhalter, D., Thomas, A., and Best, N. (2009). The BUGS project: Evolution, critique and future directions. *Statistics in Medicine*, 28(25):3049–3067.

Lunn, D. J., Thomas, A., Best, N., and Spiegelhalter, D. (2000). WinBUGS

- A Bayesian modelling framework: Concepts, structure, and extensibility. *Statistics and Computing*, 10(4):325–337.

Mäntyniemi, S., Romakkaniemi, A., and Arjas, E. (2005). Bayesian removal estimation of a population size under unequal catchability. *Canadian Journal of Fisheries and Aquatic Sciences*, 62:291–300.

Maunder, M. N. and Punt, A. E. (2013). A review of integrated analysis in fisheries stock assessment. *Fisheries Research*, 142:61–74.

McCarthy, M. A. (2007). *Bayesian Methods for Ecology*. Cambridge University Press, Cambridge, UK, 1st edition.

McCarthy, M. A. and Masters, P. (2005). Profiting from prior information in Bayesian analyses of ecological data. *Journal of Applied Ecology*, 42(6):1012–1019.

Methot, Jr., R. D. and Wetzell, C. R. (2013). Stock synthesis: A biological and statistical framework for fish stock assessment and fishery management. *Fisheries Research*, 142:86–99.

Meyer, R. and Millar, R. B. (1999). BUGS in Bayesian stock assessments. *Canadian Journal of Fisheries and Aquatic Sciences*, 56(6):1078–1087.

Millar, R. B. and Meyer, R. (2000). Bayesian state-space modeling of age-structured data: Fitting a model is just the beginning. *Canadian Journal of Fisheries and Aquatic Sciences*, 57(1):43–50.

Nelson, B. and Quinn II, T. J. (1987). Population parameters for rougheye rockfish (Sebastes aleutianus). *Proceedings of the International Rockfish Symposium*.

Ogle, D. H. (2016). *Introductory Fisheries Analyses with R*. CRC Press, Boca Raton, Florida, 1st edition.

Otis, D. L., Burnham, K. P., White, G. C., and Anderson, D. R. (1978). Statistical Inference from Capture Data on Closed Animal Populations. *Wildlife Monographs*, 62:3–135.

Plummer, M. (2003). JAGS: A program for analysis of Bayesian graphical models using Gibbs sampling. *Proceedings of the 3rd International Workshop on Distributed Statistical Computing*.

Plummer, M. (2024). *rjags: Bayesian Graphical Models using MCMC*. R package version 4-16.

Plummer, M., Best, N., Cowles, K., Vines, K., Sarkar, D., Bates, D., Almond, R., and Magnusson, A. (2024). *coda: Output Analysis and Diagnostics for MCMC*. R package version 0.19-4.1.

Pollock, K. and Pine, W. (2007). The design and analysis of field studies to estimate catch-and-release mortality. *Fisheries Management and Ecology*, 14:123–130.

Pollock, K. H., Hoenig, J. M., Hearn, W. S., and Calingaert, B. (2001). Tag reporting rate estimation: 1. An evaluation of the high-reward tagging method. *North American Journal of Fisheries Management*, 21(3):521–532.

Quinn, T. J. and Deriso, R. B. (1999). *Quantitative Fish Dynamics*. Oxford University Press, Oxford, New York.

R Core Team (2024). *R: A Language and Environment for Statistical Computing*. R Foundation for Statistical Computing, Vienna, Austria.

Regehr, E. V., Hostetter, N. J., Wilson, R. R., Rode, K. D., Martin, M. S., and Converse, S. J. (2018). Integrated population modeling provides the first empirical estimates of vital rates and abundance for polar bears in the Chukchi Sea. *Scientific Reports*, 8:16780.

Ricker, W. E. (1958). *Handbook of Computations for Biological Statistics of Fish Populations*, volume 119. Fisheries Research Board of Canada, Ottawa.

Ricker, W. E. (1975). *Computation and Interpretation of Biological Statistics of Fish Populations*, volume 191. Dept. of Fisheries and Oceans : Minister of Supply and Services Canada, Ottawa.

Robson, D. S. and Regier, H. A. (1964). Sample size in Petersen mark–recapture experiments. *Transactions of the American Fisheries Society*, 93(3):215–226.

Royle, J. A. (2004). N-Mixture models for estimating population size from spatially replicated counts. *Biometrics*, 60(1):108–115.

Royle, J. A. (2008). Modeling individual effects in the Cormack–Jolly–Seber model: A state–space formulation. *Biometrics*, 64(2):364–370.

Royle, J. A. and Dorazio, R. M. (2008). *Hierarchical Modeling and Inference in Ecology: The Analysis of Data from Populations, Metapopulations and Communities*. Academic Press, Amsterdam, 1st edition.

Schaub, M., Maunder, M. N., Kéry, M., Thorson, J. T., Jacobson, E. K., and Punt, A. E. (2024). Lessons to be learned by comparing integrated fisheries stock assessment models (SAMs) with integrated population models (IPMs). *Fisheries Research*, 272:106925.

Scherrer, S. R., Kobayashi, D. R., Weng, K. C., Okamoto, H. Y., Oishi, F. G., and Franklin, E. C. (2021). Estimation of growth parameters integrating tag-recapture, length-frequency, and direct aging data using likelihood and Bayesian methods for the tropical deepwater snapper Pristipomoides filamentosus in Hawaii. *Fisheries Research*, 233:105753.

Spiegelhalter, D. J., Best, N. G., Carlin, B. P., and Van Der Linde, A. (2002).

Bayesian measures of model complexity and fit. *Journal of the Royal Statistical Society. Series B, Statistical Methodology*, 64(4):583–639.

Štrumbelj, E., Bouchard-Côté, A., Corander, J., Gelman, A., Rue, H., Murray, L., Pesonen, H., Plummer, M., and Vehtari, A. (2024). Past, present and future of software for Bayesian inference. *Statistical Science*, 39(1):46–61.

Su, Y.-S. and Yajima, M. (2024). *R2jags: Using R to Run JAGS*. R package version 0.8-9.

Then, A. Y., Hoenig, J. M., Hall, N. G., Hewitt, D. A., and Handling editor: Ernesto Jardim (2015). Evaluating the predictive performance of empirical estimators of natural mortality rate using information on over 200 fish species. *ICES Journal of Marine Science*, 72(1):82–92.

Vetter, E. (1988). Estimation of natural mortality in fish stocks - a review. *Fishery Bulletin*, 86(1):25–43.

Wang, Y.-G., Thomas, M. R., and Somers, I. F. (1995). A maximum likelihood approach for estimating growth from tag–recapture data. *Canadian Journal of Fisheries and Aquatic Sciences*, 52(2):252–259.

Watanabe, S. (2010). Asymptotic equivalence of Bayes cross validation and widely applicable information criterion in singular learning theory. *J. Mach. Learn. Res.*, 11:3571–3594.

White, G. C., Anderson, D. R., Burnham, K. P., and Otis, D. L. (1982). Capture-recapture and removal methods for sampling closed populations. Technical report, Los Alamos National Laboratory.

Williams, B. K., Nichols, J. D., and Conroy, M. J. (2002). *Analysis and Management of Animal Populations*. Academic Press, Chantilly, 1st edition.

Winker, H., Carvalho, F., and Kapur, M. (2018). JABBA: Just Another Bayesian Biomass Assessment. *Fisheries Research*, 204:275–288.

Xie, Y. (2025a). *bookdown: Authoring Books and Technical Documents with R Markdown*. R package version 0.42.

Xie, Y. (2025b). *knitr: A General-Purpose Package for Dynamic Report Generation in R*. R package version 1.50.

Zhang, Z., Lessard, J., and Campbell, A. (2009). Use of Bayesian hierarchical models to estimate northern abalone, Haliotis kamtschatkana, growth parameters from tag-recapture data. *Fisheries Research*, 95(2):289–295.

Index

For Product Safety Concerns and Information please contact our EU
representative GPSR@taylorandfrancis.com
Taylor & Francis Verlag GmbH, Kaufingerstraße 24, 80331 München, Germany

www.ingramcontent.com/pod-product-compliance
Lightning Source LLC
Chambersburg PA
CBHW070716220326
41598CB00024BA/3189